Results and Problems in Cell Differentiation

A Series of Topical Volumes in Developmental Biology

15

Editors

W. Hennig, Nijmegen and U. Scheer, Würzburg

Results and Problems in Cell Differentiation

Volume 1 · H. Ursprung (Ed)
The Stability of the Differentiated State

Volume 2 · J. Reinert; H. Ursprung (Eds)
Origin and Continuity of Cell Organelles

Volume 3 · H. Ursprung (Ed)
Nucleic Acid Hybridization in the Study of Cell Differentiation

Volume 4 · W. Beermann (Ed)
Developmental Studies on Giant Chromosomes

Volume 5 · H. Ursprung; R. Nöthiger (Eds)
The Biology of Imaginal Disks

Volume 6 · W. J. Dickinson; D. T. Sullivan
Gene-Enzyme Systems in Drosophila

Volume 7 · J. Reinert; H. Holtzer (Eds)
Cell Cycle and Cell Differentiation

Volume 8 · W. Beermann (Ed)
Biochemical Differentiation in Insect Glands

Volume 9 · W. J. Gehring (Ed)
Genetic Mosaics and Cell Differentiation

Volume 10 · J. Reinert (Ed)
Chloroplasts

Volume 11 · R. G. McKinnell; M. A. DiBerardino
M. Blumenfeld; R. D. Bergad (Eds)
Differentiation and Neoplasia

Volume 12 · J. Reinert; H. Binding (Eds)
Differentiation of Protoplasts and of Transformed Plant Cells

Volume 13 · W. Hennig (Ed)
Germ Line – Soma Differentiation

Volume 14 · W. Hennig (Ed)
Structure and Function of Eukaryotic Chromosomes

Volume 15 · W. Hennig (Ed)
Spermatogenesis: Genetic Aspects

Spermatogenesis
Genetic Aspects

Edited by W. Hennig

With 25 Figures

Springer-Verlag Berlin Heidelberg GmbH

Professor Dr. WOLFGANG HENNIG
University of Nijmegen, Department of Genetics
Faculty of Sciences
Toernooiveld, 6525 ED Nijmegen
The Netherlands

ISBN 978-3-662-22423-6   ISBN 978-3-540-47184-4 (eBook)
DOI 10.1007/978-3-540-47184-4

Library of Congress Cataloging in Publication Data. Spermatogenesis: genetic aspects. (Results and problems in cell differentiation 15); Includes bibliographies and index. 1. Spermatogenesis – Genetic aspects. 2. Cellular control mechanisms. I. Hennig, Wolfgang, 1941–. II. Series. QH607.R4   vol. 15   574.87'612 s   87-20518 [QL966]   [591.1'66]

This work is subject to copyright. All rights are reserved, whether the whole or part of the material is concerned, specifically the rights of translation, reprinting, re-use of illustrations, recitation, broadcasting, reproduction on microfilms or in other ways, and storage in data banks. Duplication of this publication or parts thereof is only perm tted under the provisions of the German Copyright Law of September 9, 1965, in its version of June 24, 1985, and a copyright fee must always be paid. Violations fall under the prosecution act of the German Copyright Law.

© Springer-Verlag Berlin Heidelberg 1987
Originally published by Springer-Verlag Berlin Heidelberg New York in 1987
Softcover reprint of the hardcover 1st edition 1987

The use of registered names, trademarks, etc. in this publication does not imply, even in the absence of a specific statement, that such names are exempt from the relevant protective laws and regulations and therefore free for general use.

2131/3130-543210

Preface

Spermatogenesis is one of the fundamental but at the same time also one of the most complex differentiation processes in higher eukaryotes. For a long time the development of spermatozoa has been considered as related solely to the needs of transfer of the paternal genome into the egg. Other paternal contributions to the development of the embryo were not seriously considered. Only recently has it become evident, from studies of mouse embryonic development by Solter and colleagues, that also the paternal genome carries regulatory information into the zygote since the development of a mouse embryo requires the presence of the maternal as well as of the paternal pronucleus. This means that we have to pay more attention to the development of male gametes, since the imprinting of the paternal genome obviously required for the early embryonic development must occur during male gamete development.

Despite the fundamental character of sperm development as a cellular differentiation process, no coherent concepts for studies of this process exist. Many morphological, and in particular ultrastructural, details of sperm development and sperm structure are known, but this knowledge has not been assembled into a consistent picture reflecting the basic features of this differentiation process. One of the reasons for the failure to construct such a picture is the fact that also the genetics of sperm development is poorly developed. For elaborating concepts of early embryonic development the study of mutants has been, and still is, indispensible. However, genetic experimentation related to spermatogenesis is much more difficult, since major morphogenetic steps occur by regulation at levels other than transcription. This makes it difficult or even impossible to trace the initial defects on the basis of a clearcut and informative phenotype. As a consequence of these problems, only few systems will be suited to study the genetics of spermatogenesis successfully.

The aim of this volume is to collect the information available on the genetics of spermatogenesis from the three major animals accessible to such an experimental approach, mice, *Drosophila,* and *Caenorhabditis*. Naturally, the focus of the genetics in these three systems differs, due to the special features of each system. While the genetics of mouse spermatogenesis can only be studied by using accidentally available mutations connected with male sterility, genetics in *Drosophila* might be expected to be more directed. This state has, however, not yet been achieved. The special features of *Caenorhabditis,* in particular hermaphroditism, raise special questions, such as the mechanism of sex determination in germ cells.

I hope that this volume will help to induce more systematic approaches towards experimental studies of spermatogenesis. The major critical questions have been treated in the different chapters and the actual state of knowledge has been assembled in a manner which should be helpful for new approaches. I highly appreciate the efforts of the authors and their cooperation.

Nijmegen, July 1987 WOLFGANG HENNIG

Contents

Genetic Control of Spermatogenesis in Mice
By M. A. Handel (With 5 Figures)

1	Introduction	1
1.1	Description of Spermatogenesis in the Mouse	1
1.1.1	Spermatogonia	2
1.1.2	Meiosis	3
1.1.3	Spermiogenesis	4
2	Genetic Conditions Affecting Spermatogenesis	4
2.1	Defined Genes Affecting Spermatogenesis and Sperm Structure	4
2.1.1	Mutations Affecting Spermatogenesis: An Annotated List	6
2.2	Genetic Control of Sperm Morphology	12
2.3	Genetic Control of Sperm Function	14
2.3.1	The *t* Haplotypes: Background and General Effects	14
2.3.2	Effects of the *t* Complex on Sperm Development and Function	15
2.3.2.1	Transmission Ratio Distortion	15
2.3.2.2	Sterility of Males with *t* Haplotypes	18
2.4	Do Genes Affecting Spermatogenesis Act Autonomously Within Germ Cells?	20
2.5	"Male-Fertility" Genes vs "Spermatogenesis" Genes	21
3	Chromosome Anomalies Affecting Spermatogenesis	23
3.1	Sex-Chromosome Anomalies	23
3.1.1	Sex-Chromosome Aneuploidy	23
3.1.2	Sex-Chromosome Translocations	24
3.2	Autosomal Chromosome Anomalies	27
3.2.1	Characteristics of Male-Sterile Autosome Translocations	27
3.3	Hypotheses on Causes of Chromosome Sterility	31
4	Structure and Role of the Sex Chromosomes in Spermatogenesis	34
4.1	The Sex Vesicle	35
4.2	Allocyclic Behavior of the Sex Chromosomes	37
4.3	X-Chromosome Inactivation as a Correlate of Spermatogenesis	38
4.4	Role of the Y Chromosome in Spermatogenesis	39
5	Overview of Molecular Events of Spermatogenesis	41
5.1	Spermatogonial Stages	41
5.2	Spermatogenic Meiosis	42
5.2.1	DNA Synthesis During Meiosis	42

5.2.2	RNA Transcription During Meiosis	42
5.3	Spermiogenesis	43
5.3.1	RNA Synthesis During Spermiogenesis	44
6	Haploid Gene Action During Spermatogenesis	44
6.1	Postmeiotic Synthesis of RNA	45
6.2	Molecular Evidence for Transcription of Genes During the Haploid Phase	45
6.2.1	Specific Genes Transcribed During the Haploid Phase	46
6.3	How Significant is Haploid Gene Action During Spermatogenesis?	48
6.3.1	Cytoplasmic Bridges	48
6.3.2	Time of Gene Action	48
6.3.3	Genetic Evidence	50
7	Resumé and Prospectus	51
References		52

Spermatogenesis in *Drosophila*

By J. H. P. HACKSTEIN (With 18 Figures)

1	Introduction	63
1.1	A Short Description of Spermatogenesis	63
2	The Genetic Requirements for Spermatogenesis	66
2.1	The Number of Genes Affecting Spermatogenesis	66
2.2	Chromosome Rearrangements Causing Male Sterility	67
3	The Y Chromosome	68
3.1	Functions of the Y Chromosome	68
3.2	The Genetics of the Y Chromosome of *D. melanogaster*	68
3.3	The Cytogenetics of the Y Chromosome of *D. melanogaster*	77
3.4	The Genetics of the Y Chromosome of *D. hydei*	83
3.5	The Cytogenetics of the Y Chromosome of *D. hydei*	89
3.6	Lampbrush Loops Formed by the Y Chromosome of *D. hydei* and *D. melanogaster*	93
3.7	Y Chromosome Mutations and Spermatogenesis	97
3.8	DNA Sequences from the Y Chromosome of *D. hydei*	102
4	Conclusions	105
References		108

Genetic Control of Sex Determination in the Germ Line of *C. elegans*

By J. KIMBLE (With 2 Figures)

1	Introduction	117
2	Background	117
2.1	The Organism, *C. elegans*	117
2.2	Development and Anatomy of the Germ Line	119
2.3	The Global Sex Determination Genes	120

3	Control of the Sperm/Oocyte Decision	122
3.1	Influence of Cell Ancestry on the Sperm/Oocyte Decision	122
3.2	Influence of Chromosome Sex on the Sperm/Oocyte Decision	123
3.3	Influence of the Soma on the Sperm/Oocyte Decision	123
3.4	Genetics of Sex Determination in the Germ Line Tissue	124
3.4.1	Selections and Screens for Mutations that Sexually Transform the Germ Line	124
3.4.2	Global Sex Determination Genes with Germ Line-Specific Controls	124
3.4.3	Germ Line-Specific Sex Determination Genes	125
4	Conclusions	126
References		127

Subject Index 129

Genetic Control of Spermatogenesis in Mice

Mary Ann Handel[1]

1 Introduction

In this chapter I shall review evidence for direct genetic control of spermatogenesis in mice from two perspectives. First, I shall consider genetic conditions affecting the process of spermatogenesis: both specific gene mutations and chromosome anomalies causing interruption of the spermatogenic process. Second, I shall discuss the molecular evidence for direct gene action during spermatogenesis, and for changes in the patterns of gene and chromosome activity during spermatogenesis that might be implicated in controlling events.

Because the focus of this review is on the genetic control of the immediate events of spermatogenesis, I will not cover a variety of mutations controlling aspects of sex differentiation and reproductive function that are also related to the process of spermatogenesis. These include mutations affecting primary sex differentiation (see Eicher and Washburn 1986), mutations causing germ-cell depletion in both sexes, and genetic endocrinological defects affecting both sexes (see both Searle 1982; Chubb and Nolan 1985 for reviews of mutations in these two categories).

1.1 Description of Spermatogenesis in the Mouse

The morphological events of mammalian spermatogenesis have been reviewed previously (Fawcett 1975; Bellve 1979; Bellve 1982; Bellve and O'Brien 1983). However, a brief recapitulation here will be useful as a foundation for the consideration of the genetic control of the process.

The complex and highly integrated events comprising spermatogenesis include mitotic proliferation of the spermatogonial cells that are connected by intercellular bridges. These form primary spermatocytes that undergo meiosis. The haploid products of meiosis differentiate into streamlined and morphologically unique spermatozoa by a process called spermiogenesis. These events occur within the seminiferous tubules of the testis, which contain spermatogenic cells and a single type of somatic cell, the Sertoli cell (Fig. 1).

---

[1] Department of Zoology and Graduate Program in Cellular, Molecular, and Developmental Biology, University of Tennessee, Knoxville, Tennessee 37996-0810, USA.

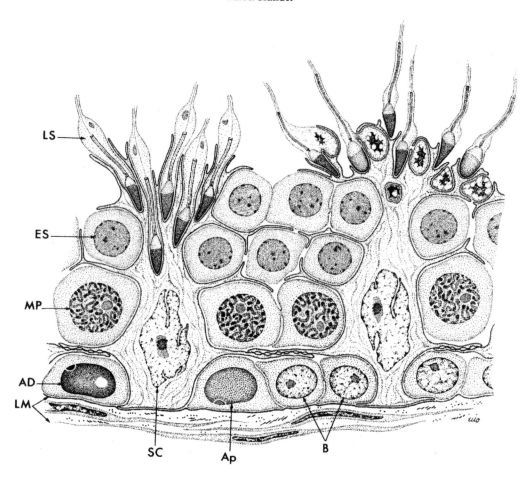

Fig. 1. The mammalian seminiferous epithelium showing relationships of the germ cells to the Sertoli cells. *SC* Sertoli cell; *Ap* type A pale spermatogonium; *AD* type A dark spermatogonium; *B* type B spermatogonium; *MP* middle pachytene spermatocyte; *ES* early (round) spermatid; *LS* late (elongating and condensed) spermatids; *LM* limiting membrane of the seminiferous tubule. (Courtesy of Y. Clermont)

1.1.1 Spermatogonia

Type A and B spermatogonia are located adjacent to the basement membrane of the seminiferous tubules. The mitotic kinetics of these cells are complex, and the identity of the renewing Type A stem cell is still controversial (for review and literature citations, see Bellve 1979). After a finite number of mitoses, Type B spermatogonia give rise to primary spermatocytes (the preleptotene spermatocytes). These leave the mitotic population, undergo a terminal round of DNA replication, and enter meiotic prophase.

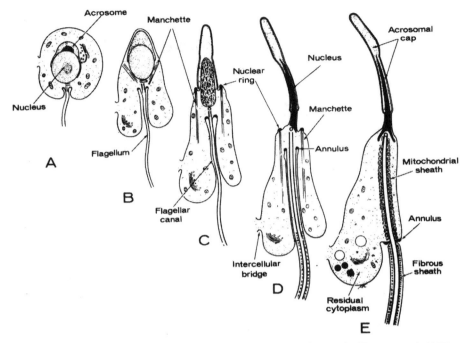

Fig. 2 A–E. The major morphological events of mammalian spermiogenesis. (Fawcett et al. 1971)

1.1.2 Meiosis

Coincident with early meiotic function, spermatocytes move away from the basement membrane of the seminiferous tubule, passing through the junctions of the Sertoli cells from the basal compartment to the adluminal compartment of the seminiferous tubule. The adluminal compartment is a microenvironment sequestered from the body by the blood-testis barrier, the morphological basis of which is the specialized Sertoli cell junctions. (For a review of the relationships between germ and Sertoli cells, see Russell 1980.) The effect of this barrier is that spermatogenic cells are relatively inaccessible physiologically, and that their unique antigens are not recognized by the immune system.

Primary spermatocytes undergo the lengthy and complex processes of meiotic prophase, involving pairing of homologous chromosomes, repair synthesis of DNA presumably associated with recombination, and extensive RNA synthesis. Pachytene spermatocytes are the largest of all the spermatogenic cells, and this stage is the longest phage of spermatogenesis (11 of 36 days in the mouse). The first meiotic division accomplishes segregation of homologous chromosomes, producing secondary spermatocytes. These cells rapidly undergo the second meiotic division to produce haploid, round spermatids.

1.1.3 Spermiogenesis

The final phase of spermatogenesis, spermiogenesis, is remarkable for the intricate and beautiful cytological processes that result in the highly differentiated spermatozoon (Fig. 2). The round nucleus undergoes elongation and the chromatin becomes tightly compacted and condensed, with concomitant changes in the repertoire of chromosomal proteins. The nucleus of mouse spermatids is flattened and acquires a rostral curvature, resulting in the species- (and sometimes strain-) specific shape. During spermiogenesis, the nucleus becomes invested by a membranous derivative of the Golgi apparatus, the acrosome. This structure contains enzymes that will subsequently digest a path for the sperm through the outer vestments of the egg.

The complex tail formed during spermiogenesis is characterized by the typical eukaryotic axonemal complex of two inner singlet and nine outer doublet microtubules. Throughout much of its length, this axonemal complex is surrounded by nine outer dense fibers. These structures are unique to mammalian sperm cells; their function is not understood. Peripheral to the outer dense fibers the mitochondria form a helical coil in the anterior portion (or middle piece) of the tail. Another unique structure, the fibrous sheath, develops in the principal piece of the tail. Amazingly, almost nothing is known of the mechanisms controlling the synthesis and assembly of these specialized components of the mouse sperm tail.

2 Genetic Conditions Affecting Spermatogenesis

2.1 Defined Genes Affecting Spermatogenesis and Sperm Structure

It is intellectually satisfying and scientifically informative to provide a genetic dissection of a process, leading ultimately to an understanding of gene hierarchies that order events of a developmental program. For analysis of the progression of mammalian spermatogenesis, one would like a series of mutations affecting spermatogenesis at various points, in such a way that one could order a series of determinative events. Unfortunately, such a series of genes has not been described in the mammalian species. In fact, there has been no directed effort to discover genes whose sole effect is on spermatogenesis. Problems associated with such a program are formidable and daunting with respect to both expense and time. Screening for male-sterile genes would involve lengthy breeding testing to detect affected males (whose only distinguishing phenotype would be failure to produce offspring) and also exhaustive test crosses with subsequent breeding tests to identify the heterozygotes. The magnitude of this effort has been sufficient to discourage any attempts at screening for male sterility. As a consequence, those mutations which have been discovered to affect sperm production are pleiotropic, with primary phenotypic effects other than sterility. These other effects not only have provided the first identification of the mutant, but also have subsequently been used as a marker in breeding programs.

Table 1. Mouse gene mutations affecting male fertility[a]

Gene symbol	Gene name	Effects
bc	Bouncy	Normal sperm count; 4.8% abnormal sperm[b]
bs	Blind-sterile	Abnormal spermiogenesis (Sotomayor and Handel 1986)
c^{3H}/c^{6H}	Albino-deletion heterozygotes	Abnormal spermiogenesis (Lewis et al. 1978)
cn	Achondroplasia	Normal sperm count; 23.9% abnormal sperm[b]
fl	Flipper-arm	Reduced sperm count (Chubb and Nolan 1985)
ho^{2J}	Hot-foot	Normal sperm count; 24.8% abnormal sperm[b]
hop (and hop^{hpy})	Hop-sterile	Abnormal spermiogenesis (Johnson and Hunt 1971; Bryan 1977b)
Hst-1, Hst-2	Hybrid sterility	Causes failure in spermatogenesis in certain interspecific hybrids (Forejt and Ivanyi 1975; Bonhomme et al. 1982)
Hx	Hemimelic extra toes	Normal sperm count; 11.4% abnormal sperm[b]
jg	Jagged tail	Reduced viability, small testes (Green 1964; 1981)
lst	Strong's luxoid	Apparently normal spermatogenesis (Forsthoefel 1962)
lu	Luxoid	Absence of spermatogonia (Elkins and Forsthoefel, cited in Johnson and Hunt 1971)
mh	Mocha	Normal sperm count; 11.4% abnormal sperm[b]
myd	Myodystrophy	Sperm count 20% normal; 17.5% abnormal sperm[b]
olt	Oligotriche	Azoospermia (Moutier 1976)
p^{6H}, p^{25H}, p^S	Pink-eyed, sterile	Abnormal spermiogenesis (Hunt and Johnson 1971; Bryan 1977a)
pcd	Purkinje-cell degeneration	Abnormal spermiogenesis (Handel and Dawson 1981)
qk	Quaking	Abnormal spermiogenesis (Bennett et al. 1971)
spa	Spastic	Normal sperm count; 9.0% abnormal sperm[b]
stb	Stubby	Sperm count and spermatogenesis unaffected (Chubb and Nolan 1985)
Various t haplotypes		See text
wr	Wobbler	Abnormal spermiogenesis (Leetsma and Sepsenwol 1980)
Xpl	X-linked polydactyly	Cryptorchid testes

[a] All information not otherwise referenced pertains to unpublished observations of Handel and Dawson.
[b] Not statistically exciting.

Mutations rendering males but not females sterile are good candidates for genes affecting the process of spermatogenesis. A comprehensive list of such mutations is provided in Table 1. We have screened a number for the possible presence of sperm abnormalities. Sperm were expressed from the epididymis and vas deferens of both homozygous affected males and unaffected littermates, and scored for the following parameters: sperm number, sperm motility, and sperm morphology (Wyrobeck and Bruce 1975, 1978). In this manner a significant proportion of these mutations was eliminated as apparently not affecting spermatogenesis (see Table 1) but probably exerting their effect on male fertility in other ways (such as by preventing mating because of limb deformities or olfactory disturbances). Those mutations that affect sperm morphology are described in detail below. They are all, as previously mentioned, pleiotropic with phenotypic effects other than male sterility. In no case is it known if the effect on spermatogenesis is a direct consequence of gene action (or failure in gene action) or if it is an indirect one. By and large, these mutations are not easily amenable to experimental analyses. In practically all cases, the testes are small; in many instances, they show varying degrees of germ-cell depletion and the adult males are rendered virtually azoospermic. These facts mean that biochemical analyses, which in normal mice are facilitated by modern methods of testis cell separation (such as the STAPUT technique, see Bellve et al. 1977), are almost entirely precluded. Also difficult, if not impossible, are analyses of sperm function, either in vivo or in vitro. Nonetheless, as will be shown, study of these mutants has provided both descriptive and experimental information that has been useful in defining some parameters of the process of spermatogenesis and spermiogenic morphogenesis.

2.1.1 Mutations Affecting Spermatogenesis: An Annotated List

azh. Abnormal spermatozoon head (*azh*) is an exception in this list of mutations in that it affects sperm morphology but does not cause male sterility. Although the mutation is an autosomal recessive, linkage has not been demonstrated. This interesting mutation, described by Hugenholtz (1984), causes abnormally shaped sperm heads and apparently no other effects. The abnormalities in sperm-head shape do not seem to affect sperm function since the homozygous mice are fertile.

bs. Blind-sterile (*bs*) is an autosomal recessive mutation (Chromosome 2) described originally by Varnum (1983). Homozygous recessive individuals of both sexes are characterized by bilenticular cataracts, with the eyes slightly smaller than those of animals without the cataracts. The coat of affected individuals is slightly glossy. Females are fertile and raise litters; males are totally sterile.

Parameters of sterility in the males have been described by Sotomayor and Handel (1986). Males copulate, as determined by the presence of vaginal plugs in females who have been caged overnight with mutant males; however, these matings are sterile. The accessory glands of these males appear grossly normal. The epididymides and vasa deferentia are virtually devoid of spermatozoa. Testes of affected males are small and are relatively, but not totally, germ-cell deficient, with less than 50% of the seminiferous tubule cross-sections containing spermiogenic stages. Since germ-cell deficiency was detected in 5-day-old mutant males (Sotomayor et al. 1986) it appears to be a developmental rather than aging phenomenon. In spite of germ-cell deficiency, both ultrastructural and cytogenetic analyses revealed normal appearance of spermatogonia and of primary spermatocytes.

The most striking effect of the *bs* mutation on spermatogenesis is the complete absence of acrosomes in differentiating spermatids. This was initially detected by employing a periodic acid-Schiff (PAS) stain (Leblond and Clermont 1952; Oakberg 1956). Although PAS-positive granules were seen in the cytoplasm of early round spermatids in *bs/bs* mice, there was no staining of the acrosomal region of any spermatids from the testes of mutant males. Ultrastructural analyses revealed the formation of proacrosomal vesicles in these spermatids, the attachment of multiple vesicles to the nuclear envelope, but no further differentiation of acrosomes. An intriguing observation was that the nuclear envelope subjacent to the attached proacrosomal vesicles in these spermatids underwent the characteristic thickening observed in normal spermatids. This thickening of the nuclear envelope spread in a caudal direction just as though an acrosome were accomplishing its caudal growth, although in fact, in no case was there an acrosome. Those spermatids that did progress to later stages of development exhibited both nuclear elongation and chromatin condensation. This observation for the first time permits the inference that the processes of nuclear elongation and chromatin condensation are not dependent in a mechanistic sense on acrosome differentiation. Since head shape in *bs* spermatids is highly aberrant, a role for the acrosome in determining final head shape cannot be excluded. However, it should be noted that since abnormal sperm-head shape is a characteristic of all male-sterile mutants, many of which have no acrosome abnormalities, the occurrence of abnormal sperm heads in the acrosome-less *bs* spermatids does not necessarily implicate the acrosome in sperm-head shaping.

c. The albino (*c*) locus is on Chromosome 7 and affects the amount of tyrosinase in pigment cells (Coleman 1962). Although alleles at this locus are not usually associated with any impairment of spermatogenesis, it has been found that mice doubly heterozygous for two radiation-induced alleles (c^{3H} and c^{6H}) are viable but runted and sterile (Lewis et al. 1978). Female oogenesis is normal but nearly all of the fetuses die. Spermatogenesis of the c^{3H}/c^{6H} males is directly affected. Epididymal sperm have abnormal head morphology and are nonmotile. The numbers of primary spermatocytes and differentiating spermatids within the testes are reduced. Spermatids in late stages of differentiation manifest head-shape abnormalities and vacuoles within the heads. Since these radiation-induced alleles are small deletions, it is not known if the effects on spermatogenesis are due to failure in action of the *c* locus or to a mutation in another, closely linked gene.

hop and hop^{hpy}. Hop-sterile (*hop*) was originally described by Johnson and Hunt (1971) and hydrocephalic-polydactyl (*hpy*) by Hollander (1976). These two mutations have now been demonstrated to be allelic (Handel 1985). The mutation is autosomal recessive, and *hpy* is located on Chromosome 6 (Hollander 1976). Expression of the mutant gene causes preaxial polydactyly of both the hind and fore feet; the afflicted mice have a rabbitlike hopping gait whereby both hind legs are used simultaneously. Mutant mice are frequently characterized by hydrocephaly, poor growth, and reduced viability, although these conditions are expressed to a lesser extent on a hybrid background (Handel and Park, unpublished). Mutant females are fertile; mutant males are sterile.

Aspects of spermatogenesis in mutant males have been described for both the *hop* and *hop^{hpy}* alleles (Johnson and Hunt 1971; Bryan 1977b, respectively). The most striking feature is the complete absence of structurally complete sperm tails in these mutant mice. Spermatid centrioles appear normal and are in the appropriate position relative to the nucleus. There is some abortive tail development, since occasionally loose groupings of outer dense fibers and mitochondria are observed. Both Johnson and Hunt (1971) and Bryan (1977b) described dense staining material within the spermatid tail which could be either aberrant doublets or outer dense fibers. It appears that spermatid tail growth is either not extensive or is followed by disorganization and degradation. However, it is unclear whether the failure of spermatid tail formation in these mutant mice is due to cessation of the entire process of spermiogenesis or can be attributed to a specific lack of molecular precursors for axonemal structures. The formation of normal-appearing tracheal cilia in these animals (Bryan and Chandler 1978) may be construed as an argument against the latter alternative, unless, of course, there is a lack of testis-specific forms of the molecular components of the axoneme (see Distel et al. 1984 for evidence for a testis-specific α-tubulin).

The nuclei of spermatids in homozygotes for *hop* or *hop^{hpy}*, similar to those of other male-sterile mutants, are characterized by aberrant shape with distortion of both the nucleus and the acrosome, inpocketing of Sertoli cell cytoplasm into the anterior region of the nucleus, and, in some cases, extreme elongation of the spermatid nucleus.

Hst. The hybrid sterility-1 gene (*Hst-*1) does not by itself appear to affect spermatogenesis but is expressed only in males of certain interspecific hybrids, which exhibit failure in spermatogenesis (Forejt and Ivanyi 1975). The gene was detected in a series of systematic crosses between various male wild mice and female laboratory mice, primarily C57BL/10 (abbreviated B10) and C3H/Di (abbreviated C3H). The wild males could be classified into three groups after matings with B10 females: I, those producing only sterile sons; II, those producing fertile sons; and III, those producing both sterile and fertile sons. Those producing sterile sons also do so when mated to females of the following inbred strains: A/Ph, BALB/c, DBA/1, and AKR/J; however, they produce fertile sons when mated to females of the C3H, CBA/J, P/J, I/St, and F/St strains. Histological examination of testes of the sterile males revealed that few spermatogenic cells progressed beyond spermatogonial stages, and those that did reach spermatocyte stages were characterized by univalency of autosomes and a high degree of X-Y dissociation; however, there were no aberrations of individual chromosomes. Although small sample size precluded extensive measurements, there appeared to be no differences in levels of androgens and gonadotropins between the sterile and fertile progeny. When males of Group I were mated to females bearing a marked (*T*) Chromosome 17, a bimodal distribution of male progeny was obtained consistent with the hypothesis that hybrid sterility is determined by a single gene, *Hst-*1, on Chromosome 17, between *T* and H2. Mice of the C3H and B10 strains differ in the alleles at the *Hst-*1 locus. Although this gene has not been investigated further, its close linkage to the *t* region, characterized by meiotic transmission distortion (see Sect. 2.3) makes it of great interest.

A second gene causing failure of spermatogenesis in certain hybrids, *Hst-*2, has been localized on Chromosome 9, close to the *Mod-*1 gene (Bonhomme et al. 1982).

olt. Oligotriche (*olt*) is a recessive autosomal mutation causing sparse fur and azoospermia. A very brief description of the characteristics of this mutant was given by Moutier (1976). Homozygotes show retardation in the growth of the ventral coat. The females are fertile, but the males are sterile. Testes are normal in size although the epididymis is reduced in weight. Spermatogenesis was reported to be grossly normal until spermatid stages of development; however, structure of the spermatids was not described. No mature spermatozoa are found in the seminiferous tubules or in the epididymides.

Sterile p alleles. Recessive alleles at the pink-eyed dilution locus (*p*, Chromosome 7) cause a reduction in the black pigment of coat hairs, but do not affect the yellow pigment. The pigment granules are shredlike, causing a reduction in pigment but with normal color. The homozygotes for several of the alleles have pink eyes, with a reduction in or absence of the pigmentation of the retina and choroid (Green 1981; Silvers 1979). The effects of some of the alleles are pleiotropic and include small size, nervousness, lack of coordination, and sterility in addition to effects on pigmentation. The etiology of these effects is unknown; it has been suggested (Melvold 1974) that the complex of pleiotropic effects might be explained if these mutant alleles were actually small chromosome deletions.

Four alleles at this locus, all radiation-induced, p^{6H}, p^{25H}, p^{bs} (pink-black-eyed sterile), and p^s (pink sterile) cause male sterility associated with the production of abnormal spermatozoa. Hunt and Johnson (1971) described the ultrastructure of developing spermatids in mice homozygous for p^{6H} or p^{25H} and Hollander et al. (1960) and Bryan (1977a) described anomalies of spermatogenesis in p^s homozygotes. The three alleles cause similar defects and the description which follows is a composite. These mutant males are characterized by small testes (about half the normal size); the seminiferous tubules are small in diameter, and germ-cell hypoplasia is obvious. The process of spermatogenesis through meiosis is apparently normal and the earliest aberration detectable involve the formation of the acrosome in round spermatids. The Golgi apparatus is apparently in its normal array at the anterior end of the nucleus; however, sometimes a supernumerary Golgi apparatus is observed. Proacrosomal granules are secreted, forming the acrosomal vesicle and implanting at the nuclear envelope. Often, however, a duplication of the acrosomal vesicle occurs and multiple sites of acrosome-nucleus fusion result. The frequency of these acrosome abnormalities was not documented in these reports. Later stages of spermiogenesis are characterized by the formation of highly irregular spermatid heads. Although chromatin condensation occurs at the appropriate time, Bryan (1977a) noted that it is retarded in the posterior third of the nucleus. Tail structure in spermatozoa of these mutants is normal with respect to the formation of the axial filament complex, outer dense fibers, and the fibrous sheath; however, tail duplication does occur as well as occasional abnormalities of the mitochondrial sheath. Sperm recovered from the cauda epididymis and vas deferens are motile, even the "headless" ones (Bryan 1977a).

In an effort to detect evidence for haploid expression of *p* alleles in determining sperm morphology and the possibility of allelic complementation in this process, Wolfe et al. (1977) examined the frequency of various sperm abnormalities in homozygotes for various alleles and in mice that were compounds for two presumed different sterile alleles. Several conclusions may be drawn from their observations. First, frequency distributions of abnormal sperm classes were complex even for any given sterile p allele. Likewise, the frequency distribution classes for the compounds (p^{6H}/p^{bs}, p^{6H}/p^{25H}, and p^{bs}/p^{25H}) were similarly complex. There was no evidence that complementation between alleles corrected or alleviated sperm defects. There was also no evidence for bimodality in the frequency distribution of the compounds, which might be taken as evidence for haploid gene action (if each allele were expressed autonomously within the haploid spermatids). However, given the complexity of the frequency classes of abnormalities found in the homozygotes, it is doubtful that bimodality of expression could have been detected. Furthermore, since these various alleles were on different genetic backgrounds, there is no clear prediction of results obtained from combining them on new backgrounds. Therefore, these data offer no firm evidence for or against the hypothesis of haploid expression of p alleles in causing sperm abnormalities.

In a further study designed to probe the nature of gene action by the sterile p alleles, Hash and Wolfe (1979) reported that the sterile alleles p^{6H} and p^{25H} apparently affect the negative surface charges of spermatozoa. They used colloidal iron hydroxide (CIH), which binds to negative moieties (such as sialic acid on gangliosides); binding can be visualized and quantitated both by scanning and transmission electron microscopy. Normal sperm (in this study, from homozygous p^d/p^d mice) bind CIH, whereas spermatids from mice homozygous for either p^{6H} or p^{25H} do not bind CIH. It is not known if these membrane alterations are a direct or indirect effect of action at the p locus.

The etiology of the spermatozoan defects in mutants bearing sterile p alleles or the relationship of these defects to the pigmentation abnormalities is not understood. Two similar but independent studies have suggested the possibility of endocrine defects. Melvold (1974) reported that the anterior pituitaries of sterile males had a lower proportion of gonadotropin-producing cells than the anterior pituitaries of fertile males bearing nonsterile p allelles. A caveat in the interpretation of this work derives from the fact that the genetic background (MUP) of the sterile p alleles used was different from that of the majority of the fertile alleles. Johnson and Hunt (1975) also reported endocrinological deficiencies in sterile, pink-eyed mice. They found degenerating axons in the posterior pituitary of mice homozygous for the p^{25H} allele, but not in mice homozygous for other sterile alleles (p^{11H} and p^{bs}). Mice homozygous for p^{25H} also showed reduced binding capacity for estradiol-17-β and reduced thyroid-iodine binding. However, in contrast to the study by Melvold (1974), Johnson and Hunt (1975) found no changes in the structure of the anterior pituitary of mice bearing sterile p alleles, although they did not utilize a stain specific for gonadotropic cells. Both of these studies antedate extensive use of radioimmunoassay techniques for the determination of hormone levels; questions about the endocrinological status of mice bearing sterile p alleles can be answered more definitively now.

pcd. Purkinje cell degeneration (*pcd*) (Chromosome 13) is an autosomal recessive mutation causing postnatal degeneration of virtually all of the cerebellar Purkinje cells (Mullen et al. 1976; Landis and Mullen 1978). Homozygotes exhibit moderate ataxia by 15 to 18 days and there is also a slower degeneration of the photoreceptor cells of the retina and the mitral cells of the olfactory bulb. Males are sterile and females are fertile. Abnormalities of spermiogenesis in homozygous *pcd/pcd* mice have been described by Handel and Dawson (1981). In mutant males, the sperm count is reduced 100-fold and the sperm are nonmotile and characterized by morphological abnormalities of the head and/or tail. Early to mid-differentiation stages of spermiogenesis are ultrastructurally normal in the mutant mice; acrosome formation proceeds on schedule and nuclear elongation and condensation are initiated. Later stages of spermiogenesis, however, are characterized by abnormalities of both head shaping and tail assembly. Aberrantly shaped heads frequently contain inpocketings of either spermatid or Sertoli cell cytoplasm. Defects in the assembly of outer dense tail fibers include extraneous fibers in the tail as well as ectopic deposition of tail fibers in the adjacent cytoplasm.

Causes of these defects are not known. The site of gene action in causing Purkinje cell degeneration was investigated by Mullen (1977, 1978) by the construction of chimeras in which one component was homozygous *pcd/pcd* and the other normal, $+/+$. Using a cell-specific marker (differences in histochemical staining intensity for the activity of β-glucuronidase), Mullen demonstrated in these chimeras that the *pcd/pcd* Purkinje cells degenerated, while the $+/+$ Purkinje cells survived, implying that the action of the gene in causing Purkinje cell degeneration is autonomous within the Purkinje cells. Autonomy of the action of the *pcd* mutation in male germ cells was not definitively established in this study. Two male chimeras were homozygous for *pcd* in one cellular component, were sterile, and produced abnormal sperm. However, data on abnormal sperm were not quantitative and the sex of the cellular components of these chimeras was not determined. Thus, it was not possible to establish if these sperm abnormalities were a consequence of autonomous action of the *pcd* mutation within male germ cells.

qk. Quaking (*qk*, Chromosome 17) is an autosomal recessive mutation causing deficient myelinization of the central nervous system (Sidman et al. 1964) and also of the peripheral nervous system (Samorajski et al. 1970). These mutants are characterized by marked tremors that disappear when the animal is at rest or in contact with bedding; the mutant mice are also subject to seizures. Deficiency of myelin sheaths in the nervous system is related to lipid deficiencies; this subject was reviewed by Baumann et al. (1972).

Although afflicted females are fertile, homozygous *qk/qk* males are sterile and virtually azoospermic. Defects of spermiogenesis were described by Bennett et al. (1971). The testes of these males are about half their normal size, but descend normally. Early stages of spermiogenesis, including formation of the acrosome, and early stages of tail morphogenesis, are normal in mutant males. Chromatin condensation and early nuclear elongation are initiated normally and at the correct temporal stage. However, subsequent to the initiation of chromatin condensation in step-10 spermatids, multiple abnormalities are apparent. Head shaping fails to proceed normally, the nucleus is characterized by multiple vacuoles, and fingerlike extensions of the Sertoli-cell cytoplasm indent the nucleus. There is a progressive disorganization of the tail structure, with the axonemal complex splaying out, and general disarray of both the axoneme and the outer dense fibers. These defects result in total failure of final sperm-head shaping and virtual lack of production of mature spermatozoa.

The causes of abnormal spermatid differentiation in *qk/qk* mice are not known. Coniglio et al. (1975) investigated the possibility that changes in lipid and fatty acid composition of the testes might be correlated with abnormal spermatogenesis. There were no extensive differences between mutant and control mice; however, the amount of esterified cholesterol (but not of free cholesterol) is significantly higher in the testes of mutant mice, and the concentration of phosphatidyl ethanolamine is slightly decreased in the testes of the mutants. However, these small differences may well be consequences rather than causes of abnormal spermiogenesis; in fact, it has been reported that an increase in esterified cholesterol accompanies spermatogenic impairment (Johnson 1970).

t. Effects of various *t* haplotypes on sperm structure and function will be discussed in detail below (Sect. 2.3).

wr. Wobbler (*wr*) is a recessive mutation (not mapped) causing degeneration of the motor neurons of the spinal cord and the brain stem resulting in tremors and an unsteady gait (Duchen et al. 1968). Most affected mice die by 2 or 3 months of age and none have been bred. Since neither males nor females breed, it might be expected that this mutation would not specifically affect sperm; however, Leestma and Sepsenwol (1980) have described sperm-tail abnormalities in affected mutants. Spermatozoa from the epididymis and ductus deferens are not motile, and motility could not be enhanced by treatment with 1-methyl-3-isobutyl xanthine (MIX), which did enhance motility of control spermatozoa. Ultrastructural examination revealed that the sperm from mutant mice are characterized by axoneme abnormalities, most commonly the absence of the outer doublets 4 through 7 and of the associated outer dense fibers. The defects were present in higher frequency in the distal regions of the epididymis and vas deferens (frequency being 5% in testis-cell suspensions and 70% in the sperm suspensions from the vas deferens). Since testicular histology was not examined, it is not clear if the defect can be attributed solely to degeneration of an unusually unstable axoneme in the epididymis. The authors argued that the sperm axoneme alterations observed are probably not due to generalized degeneration of the sperm cells since these cells have a generally normal appearance and no other signs of degeneration.

2.2 Genetic Control of Sperm Morphology

Although specific genes have not been identified or mapped, it has been known for some time that there are differences in sperm dimensions between mice of different strains, and therefore, a genetic component in the determination of sperm shape (evidence reviewed by Beatty 1970, 1972). There appears to be an influence of the Y chromosome, not on sperm shape, but on the incidence of sperm-head abnormalities in different strains of mice (Krzanowska 1972). Identification of genes involved in the determination of sperm morphology is complicated by the fact that there are also nongenetic contributions to the variability in sperm shape (Beatty 1970). Although it has not been possible to clearly separate genetic and epigenetic effects on sperm morphology, it is known that the action of at least some of the genetic determinants of sperm morphology is autonomous to the germ cells (see Sect. 2.4).

The issue of whether any of the genes listed in Section 2.1 have a direct and specific effect on sperm structure is confused by the degree of pleiotropy exhibited by each of the mutations. An illustration of this point is provided by the *bs* (blind-sterile) mutation, that, more than any of the other male-sterile mutations, affects an organelle unique to sperm cells, namely, the acrosome. Spermatids in homozygous *bs* mice universally fail to assemble acrosomes (Sotomayor and Handel 1986; see also Sect. 2.1). However, the mutation is also highly pleiotropic (the identifying feature of the mutation is the presence of lenticular cataracts in both male and female homozygotes). Even within the testis, the action of the mutant gene is pleiotropic. In addition to the effect on acrosome assembly, there is severe germ-cell depletion of the testes. So it is impossible to determine whether the failure in acrosome formation is due to an adverse testicular environment caused by germ-cell depletion, or whether there is an adverse testicular environment resulting in both germ-cell depletion and failure of acrosome assembly, or whether the two are independent effects of the primary lesion. It is probable that one of the last two explanations is the correct one, since other mutations also result in germ-cell depletion without any inhibition of acrosome assembly.

Another finding confusing the determination of the mode of gene action in causing sperm abnormalities is that all of the mutations discussed above result in the production of sperm characterized by a variety of morphological abnormalities: all with abnormal head shape, many with acrosome abnormalities, and many with abnormalities of the tail. Therefore, it cannot be said, for instance, that the *hop* gene specifically mediates sperm-tail formation, or that the *bs* gene is an "acrosome gene." Furthermore, many of the morphological abnormalities caused by these mutations are seen also among the abnormal spermatozoa found in low frequencies in all genetically normal male mice, and among the sperm of mice treated with any of a variety of potential mutagens, carcinogens, and teratogens (see Wyrobeck and Bruce 1975, 1978). Therefore, it is not resolved whether such abnormalities among sperm from mice with a mutation causing male sterility are a direct consequence of mutant gene action within the germ cell, or whether they are simply due to an adverse physiological microenvironment of the testis (see Sect. 2.4 for a further discussion of the ramifications of this issue).

A further complicating circumstance is that all male-sterile mutants are characterized by a low sperm count as well as sperm abnormalities. It is known that morphologically abnormal sperm are disadvantaged both in transport within the female reproductive tract (Krzanowska 1974; Nestor and Handel 1984) and in binding to the surface of the egg (Krzanowska and Lorenc 1983) and to the zona pellucida (Kot and Handel, 1987). Therefore, it is not known whether mice bearing male-sterile mutations are sterile because of morphological sperm abnormalities or low sperm count or both.

2.3 Genetic Control of Sperm Function

The effects of *t* haplotype chromosomes on sperm structure and function are broad and may well be instructive for genetic control of mammalian spermatogenesis and function, as well as an aid in the understanding of the significance of haploid gene action during spermatogenesis.

2.3.1 The *t* Haplotypes: Background and General Effects

The "*t* complex" of the mouse genome is an extensive region of chromatin on Chromosome 17. Alterations in this chromatin, evident in the variety of *t* haplotypes found in wild mouse populations, affect tail length, early embryonic survival, sperm development and function, recombination frequency, and the genetic survival of particular *t* haplotypes. Genetic analyses of mice bearing different *t* haplotypes have revealed that these are not "mutations" in the usual sense, but are regions of structurally altered chromatin (see Lyon 1981; Silver 1985).

The original detection of naturally occurring *t* haplotypes was by the nature of their interaction with a dominant gene, *T*, Brachyury. When heterozygous with the wild-type allele, *T* causes a shortening of the tail; *T* being heterozygous with any *t* haplotype (*T/t*) causes a tailless mouse. Mice with a +/*t* or *t/t* (if viable) genotype have normal tails. The various *t* haplotypes have four additional properties: (1) embryonic lethality is caused by many *t* haplotypes when homozygous (Bennett 1975; Sherman and Wudl 1977; Lyon 1981). (2) Males heterozygous for a single haplotype (+/*t* or *T/t*) have the peculiar property of passing the *t* haplotype to more than 50% of their offspring, resulting in a transmission ratio distortion (TRD). (3) Males heterozygous for two complementing lethal haplotypes or homozygous for a semilethal haplotype are sterile. (4) Complete *t* haplotypes, when heterozygous with + chromatin, cause suppression of recombination of chromatin extending from the Brachyury (T) locus to the major histocompatibility complex, MHC, or H2 complex, some 15 cM distal.

It is now clear that *t* is not a simple locus and probably is not even a complex of related genes. Instead, the *t* haplotypes are a group of structurally rearranged forms of the proximal region of Chromosome 17, including the *T* locus and the entire major histocompatibility complex. This region of the genome is 12–15 cM (perhaps $20–30 \times 10^3$ kilobases) in length, or approximately 1% of the mouse genome. The genes in this region do not appear to be functionally related but are held together as haplotypes by virtue of recombination suppression within the region.

The mouse *t* chromosomes are composed of at least two separate inversions: a distal inversion resulting in the reversal of the orders of the MHC and tufted, *tf* (Artzt et al. 1982; Shin et al. 1984), and a proximal inversion that includes the *T* and *qk* loci (Silver 1985; Herrmann et al. 1986). These two inversions effectively suppress recombination within the region covered by *t* haplotypes, resulting in the maintenance of a tight linkage of the included genes in the various *t* haplotype chromosomes. Complete *t* haplotypes encompass the two inversions in the entire *t* region, while partial *t* haplotypes (produced by rare recombination events) carry only a portion of the rearrangement, with the rest of the *t* region being wild type.

2.3.2 Effects of the *t* Complex on Sperm Development and Function

Two aspects of *t* complex effects on sperm development and function are relevant to the study of genetic control of spermatogenesis: the TRD exhibited by most males heterozygous for a single *t* haplotype and the sterility of males bearing two complete complementing or semilethal *t* haplotypes. The two may be related phenomena, and are certainly instructive in the analysis of the genetic control of sperm function.

2.3.2.1 Transmission Ratio Distortion

As described above, males heterozygous for a single complete *t* haplotype have the unusual property of passing the *t* haplotype to more than 50% of their offspring, a non-Mendelian pattern of inheritance. The TRD has been postulated to be due to the action of three distorter genes (*Tcd-1, 2,* and *3*) on a responder gene, *Tcr* (Lyon 1984). The *t* form of the responder must be present and heterozygous in order for distortion to occur. The distorter loci act additively, in cis or in trans, to increase the transmission of whichever chromosome carries the *t* form of the responder locus. This theory provides a genetic explanation for the observed effects of high and low transmission of various complete and partial *t* haplotypes, but does not explain the molecular, biochemical, or physiological parameters which allow *t*-bearing sperm to be more effective than their +-bearing meiotic partners.

Genetic background can also affect the degree of TRD. A point not much emphasized in the literature is that the available *t* haplotypes are not on the same genetic background, and, furthermore, "control" animals are often not of the same genetic background as the *t*/+ animals. Recent data reveal that the effect of genetic background on *t* haplotype expression can no longer be ignored. Gummere et al. (1986) showed that TRD is profoundly affected by both the homologous Chromosome 17 and other genetic loci. In addition, there is intramale variability in the transmission ratio. Effects of background on the function of sperm have been demonstrated by Olds-Clarke, who has made an important contribution in using congenic strains in the analysis of *t*-haplotype effects on sperm function. Olds-Clarke and McCabe (1982) produced two sets of congenic mice carrying the t^{w32} haplotype: C57BL/t-^{w32}/+ and C3H-t^{w32}/+. They used these strains to investigate two parameters of sperm function: morphological abnormalities and the ability to fertilize eggs in vitro. In each case, the controls were congenic to the *t* haplotype-bearing mice. Their results clearly implicated an influence of genetic background in the expression of the t^{w32} haplotype. Although the degree of TRD by the t^{w32} haplotype was equivalent in both strains, there were significant differences in the effects of t^{w32} on the fertilization ability of sperm in vitro and on sperm-head abnormalities. These results suggest an interaction between the genes of a t haplotype and other genetic loci affecting sperm physiology and demonstrate the importance of genetic background and the use of congenic strains in assessing the effects of t haplotypes.

It has been shown by Hillman and Nadijcka (1978a, 1978b) that there are no unique defects either of spermatogenesis or of sperm structure that might explain TRD. In the case of some t haplotypes, TRD can often be reduced by timing mating to be later than normal with respect to the time of ovulation (Braden 1958). Normally, both ovulation and mating in mice occur around the midpoint of the dark cycle. Ejaculated sperm therefore age in the female tract prior to fertilization when the eggs reach the ampulla of the oviduct. When mating is delayed with respect to ovulation, the time spent by sperm in the female tract prior to fertilization is reduced and the transmission ratio is closer to the expected Mendelian value of 50%. The transmission ratio is also closer to normal when fertilization occurs in vitro (McGrath and Hillman 1980). These data on delayed matings and fertilization in vitro imply that there is no degeneration within males of sperm bearing the wild-type Chromosome 17, and also that there is differential survival and/or more effective function of the t-bearing sperm within the female reproduction tract.

The fact that TRD is not a consequence of degeneration of wild-type sperm prior to or immediately after mating has been elegantly demonstrated by Silver and Olds-Clarke (1984) with the use of a recombinant DNA probe for a region within the t haplotype to "track" sperm in both the male and female reproductive tracts. The probe used (Roehme et al. 1984) hybridized to a single restriction fragment from the normal wild-type chromosome and to a restriction fragment of a different size from Chromosome 17 of all complete t haplotypes. Silver and Olds-Clarke used this probe to measure the ratio of t-bearing sperm and wild-type sperm in the epididymis and vas deferens from males with a high TRD and also of sperm recovered from the uterine horns of females to which they were mated. Equivalent amounts of both sizes of restriction fragments were detected in each of the sperm samples analyzed, indicating no degeneration of the wild-type sperm in the male reproductive tract or in the female tract up to 90 min postejaculation.

These analyses of TRD show that both t-bearing sperm and + sperm are present in the female reproductive tract and raise the following question: are t-bearing sperm "super" sperm, or are their meiotic partners, the + sperm, defective sperm? This question was answered recently by the analysis of results from mixed inseminations of sperm from $t^{w32}/+$ males and congenic $+/+$ males (Olds-Clarke and Peitz 1985). Control experiments demonstrated that transmission distortion occurred when females were artificially inseminated with sperm from $t^{w32}/+$ males. However, the analysis of fetuses and pups produced by artificial insemination with mixed sperm populations (from $t^{w32}/+$ and $+/+$ males) gave no evidence for transmission ratio distortion. In other words, t^{w32}-bearing sperm were no more efficient in fertilizing than were the + sperm from $+/+$ males. Similar conclusions were also reached by Seitz and Bennett (1985) in a study of the progeny from male mouse chimeras with $+/+$ and $t/+$ components. These results provide good evidence that t sperm are not superior to normal sperm and that the meiotic partners of t haplotype-bearing sperm are defective. It remains to be determined how this occurs, and precisely how the +-bearing sperm are "defective."

Examination of both biochemical and physiological parameters of sperm from mice bearing a *t* haplotype can be instructive not only in the determination of the mechanism of TRD, but also in the analysis of normal sperm function in fertilization. A diverse set of *t*-encoded polypeptides present in the testis (but not restricted to the testis) have been identified by two-dimensional gel electrophoresis (Silver et al. 1983). One of these, TCP-1, is synthesized to a greater extent in the testis than in other tissues (Silver et al. 1979). Normal testes, from mice not bearing *t* chromatin, express the basic form, while a more acidic form of this protein is encoded by *t* haplotypes (Silver et al. 1979; Danska and Silver 1980). TCP-1 is closely associated with the external surface of cells, perhaps a cell-surface component of the extracellular matrix (Silver and White 1982). However, the *Tcp-1* gene has been cloned and partially sequenced (Willison et al. 1986) and the sequence shows no homology to fibronectin. Although this protein is intriguing in that it is encoded by *t* chromatin, abundantly expressed in the testis, and associated with the membrane of germ cells, its function and relation to sperm effects such as TRD is unknown.

An additional molecule implicated in TRD is galactosyl transferase (GalTase). It has been postulated that the GalTase of sperm acts as a receptor and binds to terminal N-acetylglucosamine of the zone pellucida, thus playing a pivotal role during sperm binding and fertilization (Shur and Hall 1982). Recent evidence demonstrating the inhibition of fertilization by purified GalTase and the localization of GalTase to a specific membrane domain over the sperm acrosome supports this hypothesis (Lopez et al. 1985). Interestingly, there are alterations in the activity of this enzyme in sperm from mice bearing *t* chromatin. Sperm from mice heterozygous for a *t* haplotype have increased GalTase activity (Shur and Bennett 1979) and sperm from homozygotes have even higher activity (Shur 1981). Recombinants for *t* haplotypes which no longer exhibit TRD do not have increased GalTase levels (Shur 1981). Increased GalTase activity may be due to a deficiency in a natural inhibitor of GalTase, although such a molecule has not been chemically identified (Shur and Bennett 1979). It seems probable that these enzyme activity differences may play a role in TRD; however, the relation of the enzyme activity to *t* chromatin expression and to TRD is not yet apparent.

Another approach to the determination of the mechanism of TRD has been to examine transport and motility of sperm from $t/+$ and $+/+$ males. Using congenic strains, Olds-Clarke and co-workers have made comparisons of parameters of function and physiology of sperm from $t/+$ and $+/+$ mice. In interpreting the following experiments, it is important to remember that in most cases, *t*-bearing sperm are not being compared to +-bearing sperm, but rather that the population of sperm from $t/+$ males is being compared to the population of sperm from $+/+$ males.

Epididymal sperm from mice bearing the t^{w32} haplotype have a lower net velocity (defined as a straight line swimming speed) than sperm from congenic $+/+$ mice (Tessler et al. 1981). The apparent cause of this is decreased progressiveness (i.e., the sperm deviate from the straight line between two points) of sperm from $t^{w32}/+$ mice (Olds-Clarke 1983 b). Males carrying the most complete

t haplotypes and all distal, partial t hyplotypes are characterized by nonprogressively motile sperm (Olds-Clarke 1983a), suggesting that this characteristic maps to the distal region of the t region.

These observations on nonprogressive motility were made on epididymal sperm incubated in vitro. Since TRD does not occur in vitro, it was important to determine if there are any differences in the behavior of sperm from $t/+$ and $+/+$ males recovered from the reproductive tracts of females, where TRD does occur. Sperm from $t^{w32}/+$ mice penetrate eggs sooner both in vivo (Olds-Clarke and Becker 1978) and in vitro (Olds-Clarke and Carey 1978) than sperm from congenic $+/+$ mice. Furthermore, sperm from $t^{w32}/+$ mice are transported faster in the female reproductive tract (Tessler and Olds-Clarke 1981). Using objective and quantitative measures to assess motility characteristics of sperm, Olds-Clarke (1986) found that sperm retrieved from the isthmus of the oviduct showed greater nonlinearity in their tracks than uterine sperm. Sperm from $t^{w32}/+$ males recovered from the isthmus of the oviduct exhibited this property to a greater degree than sperm from congenic $+/+$ males. The sperm exhibiting this higher degree of nonlinearity from $t^{w32}/+$ mice were swimming faster than those sperm showing nonlinearity from $+/+$ mice.

Further quantitative work is needed to define precisely the differences between motility patterns of normal sperm during capacitation and fertilization and those of t and $+$ sperm from $t/+$ mice. The precise function of nonprogressive motility of sperm from $t/+$ mice and its relationship to the processes of capacitation, transport, fertilization, and TRD remain to be determined. It is not known if all or only some of the sperm from $t^{w32}/+$ males exhibit nonprogressive motility in the oviduct since there is considerable sperm-to-sperm variability in motility patterns. These issues are germane to the issue of haploid gene expression of t chromatin. If all sperm in the population have similar behavior, it would indicate the expression of t chromatin prior to the haploid phase of sperm development. If, on the other hand, two distinct sperm populations could be demonstrated, one with a rapid acquisition of nonprogressive motility and one without, it could be considered evidence for haploid expression of t chromatin with respect to nonprogressive motility. It would still, however, remain to be demonstrated both whether the $+$-bearing or the t-bearing sperm, had peculiar motility characteristics and to determine if nonprogressive motility plays a role at all in transmission distortion. It is not known if nonprogressive motility is advantageous or detrimental to sperm. It may be that the sperm which acquire nonprogressive motility more rapidly are defective sperm, unable to fertilize effectively and thus effectively enhancing the transmission of their meiotic partners, the t-bearing sperm, to the subsequent generation.

2.3.2.2 Sterility of Males with t Haplotypes

Males heterozygous for two lethal, complementing, complete t haplotypes are sterile; males carrying different combinations of partial t haplotypes can be fertile, sterile, or semifertile. Dooher and Bennett (1977) sought an explanation for the observed sterility of these males in an analysis of ultrastructural defects of spermiogenesis and spermatozoa in t^o/t^{w32} sterile males. They found abnormalities of nuclei, associated derangements of the microtubules of the manchette, and abnormal acrosomes. Approximately 60–75% of the sperm produced by these males had abnormally short, angular, or amorphous heads. Since the two different t haplotypes studied were on different genetic backgrounds, it was not possible to provide an adequate control for these analyses. Hillman and Nadijcka (1980; Nadijcka and Hillman 1980) analyzed spermatozoal defects of complementing t-haplotype heterozygotes (t^{Lx}/t^{Ly}) to determine if they were unique, either qualitatively or quantitatively, to these heterozygous males. They compared abnormalities of spermiogenesis and spermatozoal defects in t^6/t^{w32} males to defects observed in BALB/c, T/t^6, and T/t^{w32} control males. They concluded that all males, regardless of age, genotype, and including wild type, contain the same kinds of defective spermatids; and that no unique spermiogenic or spermatozoal defects can be found in the sterile t-haplotype heterozygous males. Therefore, it is highly unlikely that the interaction of these two haplotypes (and probably of other t haplotypes) produces specific, unique abnormalities. Nonetheless, the interaction of the t haplotypes in these heterozygous males interferes in some manner with spermiogenesis to produce the constellation of abnormalities characteristic of impaired spermiogenesis.

In addition to their morphological abnormalities, sperm from sterile t-haplotype heterozygotes are unable to accomplish various processes essential to fertilization. They are not transported to the oviduct as efficiently as normal sperm (Bennett and Dunn 1967; Olds 1970). Furthermore, uterine sperm from t^{w32}/t^o males are very slow and show little progressive motility (Olds-Clarke 1986). McGrath and Hillman (1980) also determined that sperm from sterile t-haplotype heterozygotes are unable to fertilize oviducal ova in vitro even when barriers to fertilization (cumulus cells and zonae pellucidae) are selectively removed.

Lyon (1986) has postulated that sterility of the t haplotypes is due to homozygosity of three sterility factors that map to the same partial t haplotypes as do the transmission distorter genes, and may in fact, be identical to them. In the case of TRD in $t/+$ males, the Tcd (distorter) genes are thought to act on the wild-type responder in such a way as to inhibit the sperm carrying the wild-type Chromosome 17. When the Tcd genes are homozygous (in the case of sterile males heterozygous for complementing haplotypes), their effect is proposed to extend also to the t form of the responder locus, Tcr, thereby rendering all sperm nonfunctional. In other words, male sterility may be the extreme of transmission distortion. Although the mechanism by which this might occur is not clear, this theory is attractive in that it provides a relationship between TRD and male sterility of t haplotypes.

Although many questions still remain, the analysis of sperm development and function utilizing t haplotypes as genetic tools has progressed to a much more satisfying level than that achieved by utilizing other genes that affect sperm morphology and function. In pursuing the issue of TRD, we are hopefully coming closer to an understanding not only of those genetic factors that affect TRD, but also of those aspects of sperm function that are critical to the normal process of fertilization. It is one of the promises and advantages of the study of genetics that the analysis of a mutationally or otherwise genetically disturbed process will provide insights to the normal course of events. Hopefully, work on the t complex will bring additional insights into normal sperm function and into those aspects of sperm physiology that facilitate the process of fertilization. The availability of gene probes for regions of t haplotypes makes possible a molecular understanding of the effects of this disturbed region of chromatin on both sperm structure and sperm function. The effects of the t-haplotype genes on spermatogenesis and sperm function may well provide a model for genetic control of spermatogenesis.

2.4 Do Genes Affecting Spermatogenesis Act Autonomously Within Germ Cells?

Pleiotropy of genes affecting spermatogenesis complicates the determination of the mode of gene action. It is of interest to know if genes affecting spermatogenesis are actually acting autonomously within germ cells in causing spermatogenic impairment.

This question may be analyzed by the construction of chimeric mice containing both genetically male sterile and normal cells in the testes. It is well established that the testes of male chimeras can contain two genetically different components, and, if each component is chromosomally male (XY), each can form spermatozoa (see McLaren 1976, for review). If such chimeras are constructed with one component homozygous for a gene causing male sterility and one component normal, we would expect that some of the mutant germ cells would be in proximity to normal Sertoli cells such that the hormonal and biochemical environment of the mutant germ cells might be closer or identical to that of a normal testis. If the action of the gene causing male sterility is extrinsic to the germ cell, we would expect that mutant germ cells would develop normally in a chimeric testis, resulting in offspring carrying the mutant component. Failure to rescue the mutant male sterile phenotype in the germ cells of chimeras would suggest autonomy of expression of the mutant gene within germ cells.

Such an analysis has been conducted for two different mutations affecting sperm structure: p^{6H} and qk (Handel et al. 1987). In this study, 21 male mouse chimeras were analyzed, each with chimeric testes composed of cells that were homozygous for a gene causing male sterility (either p^{6H} or qk) and of genetically normal cells. In no case did these male chimeras pass the mutant genotype to their offspring. This result demonstrates that juxtaposing normal cells to p^{6H} or qk mutant cells in chimeric testes did not rescue the male sterile phenotype of the mutant germ cells, and thereby implies autonomy of function of the p^{6H} and qk genes within spermatogenic cells.

Autonomy of gene function within mutant spermatogenic cells was also implicated by the morphological observations on the testes of these chimeric male mice. Histological and ultrastructural observations of the testes of chimeras with a homozygous p^{6H} component revealed the presence of nuclei with multiple proacrosomic vesicles, a defect characteristic of the p^{6H} phenotype, but not of the qk phenotype (and not observed in the testes of mice with a homozygous qk component). This observation is an important one for it suggests that this defect in early acrosome formation is a mutant-specific defect caused by autonomous function of the mutant allele within the germ cell and is not simply a nonspecific response to an adverse testicular environment.

Autonomy of gene function during spermatogenesis was also investigated in chimeric mice by Burgoyne (1975). As discussed above in Section 2.2, it is known that sperm shape, particularly head shape, is genetically determined (Beatty 1970, 1972). Burgoyne demonstrated that it was possible to distinguish between sperm from C57BL and C3H/Bi mice using measurements of sperm dimensions and discriminate functions. When measurements were made of sperm from male C57BL ↔C3H chimeras, the sperm clearly fell into two classes: those with a C57BL morphology and those with a C3H morphology. There was no evidence for an intermediate phenotype, such as is found in the sperm from C57BL X C3H hybrids, or for skewing of the phenotype due to the presence of genetically different cells in the testis. One of the chimeras of this analysis was an opposite sex chimera (XY ↔ XX), with a sperm phenotype characteristic of that determined by the genotype of the XY component. Within the testes of this mouse there were germ cells of only one genetic type, but Sertoli cells of both genetic types. Thus, there is no evidence from this study for a role of the Sertoli cell in the determination of these parameters of sperm shape; and the data, therefore, implicate autonomy of gene action within the germ cells.

Such autonomy of gene action in germ cells is not always the case for genes affecting male fertility. Utilizing chimeras, Lyon et al. (1975) have investigated the action during spermatogenesis of the *Tfm* (testicular feminization) gene. *Tfm* is an X-linked gene coding for an androgen receptor. The gonads of $X^{Tfm}Y$ individuals are nonresponsive to androgens, and no spermatogenesis occurs. Induction of spermatogenesis in *Tfm* germ cells would permit the determination of whether or not the androgen receptor functions on germ cells or elsewhere. Lyon et al. (1975) found that male chimeras with a *Tfm* component were fertile and bred from the $X^{Tfm}Y$ component. Therefore, the *Tfm* gene is not acting autonomously within the germ cells, and the effects of androgens on germ cells must be mediated by somatic cells (Sertoli cells?) of the testis.

These studies have demonstrated that while not all genes essential for normal spermatogenesis are acting autonomously within the germ cells, at least some genes that affect sperm morphology do act within the germ cells. The analysis of male-sterile chimeras has not brought us closer to an understanding of the primary gene effect and its role in the spermatogenic process; however, these studies have demonstrated that we should seek the causes of sterility in the events of spermatogenesis.

2.5 "Male-Fertility" Genes vs "Spermatogenesis" Genes

It is appropriate here to consider the issue of whether any "spermatogenesis" genes have been identified in the mouse. A gene should be called a spermatogenesis gene only if it can be demonstrated to specify a product or to be a controlling sequence used only during spermatogenesis. By this criterion most if not all of the genes enumerated here are not spermatogenesis genes, but are male-fertility genes. As evidenced by their pleiotropy, these genes all affect processes other than spermatogenesis as well as affecting male fertility by interfering with spermatogenesis. It is likely that these genes regulate processes or encode "housekeeping" proteins that are especially important during the differentiative phases of spermatogenesis. It is interesting to note that a significant portion of the male-sterile genes also affect the nervous system. This fact may indicate that these genes are likely to code for proteins important in the assembly of cell membranes or in cellular recognition processes (presumably also a membrane function). Spermatogenesis is a very complex and intricate developmental process and it is, therefore, not too surprising that the action of a large number of genes, not spermatogenesis genes per se, have an impact on this process.

If none of the genes acting as male-sterile genes in the mouse are spermatogenesis genes, then we must ask both why spermatogenesis genes have not yet been detected and if there are any spermatogenesis genes in the mouse. The answer to the question of why spermatogenesis genes have not been identified is probably fairly trivial: no one has looked for them. As mentioned earlier, it is both difficult and prohibitively time-consuming and expensive to screen for these mutations. It is possible that such mutations have been induced in large-scale mutagenesis programs but have gone undetected. Particularly intriguing to consider is the possibility that small deletions may include spermatogenesis genes. For example, both the p (pink-eyed dilution) and c (albino) regions are fairly well saturated with induced mutations. Some of the p alleles act as male-steriles (see above) and some the the albino deletions affect both female and male fertility (Lewis et al. 1978). Many of these mutations are small deletions and the possibility exists that the deletions cover either spermatogenesis genes or male-fertility genes.

As defined above, a spermatogenesis gene is expected either to regulate the process of spermatogenesis or to specify a protein product unique to spermatogenic cells. There is not yet any conclusive evidence for spermatogenesis regulatory genes in the mouse. However, there are assuredly at least some spermatogenesis genes specifying sperm-specific proteins. As will be discussed in a subsequent section, there are some enzyme variants unique to spermatogenic cells, and the genes encoding these sperm-specific enzymes are, by definition, spermatogenesis genes. The two best studies of these enzymes are phosphoglycerate kinase-2 (PGK-2) and lactate dehydrogenase-X, or more appropriately, lactate dehydrogenase-C_4 (LDH-C_4). The chromosome locus for LDH-C_4 is not known, but the chromosome locus for PGK-2 is Chromosome 17, adjacent to the major histocompatibility complex and the t-haplotype region (Eicher et al. 1978). Interestingly enough, this chromosome region has an unusual concentration of male-fertility and putative spermatogenesis genes: the unusual effects of the t haplotypes on the transmission ratio and male sterility suggest the possibility that they contain spermatogenesis genes; qk, in this region, is a male-fertility gene; $Hst-1$ (a gene that seems to affect only the process of spermatogenesis) is in this region; and the gene for the sperm-specific PGK-2 is in this region of chromosome 17. No other region of the mouse genome is known to have as many genes affecting spermatogenesis. Is it possible that this region of Chromosome 17 contains the mouse spermatogenesis genes?

3 Chromosome Anomalies Affecting Spermatogenesis

I have previously discussed a group of autosome mutations affecting primarily the late differentiative stages of spermatogenesis in mice. I shall now turn to a different set of genetic conditions affecting spermatogenesis: chromosome anomalies, primarily translocations. Although these chromosome translocations affect spermatogenesis more profoundly than they do oogenesis, female gametogenesis may also be impaired (Mittwoch et al. 1981; Burgoyne and Baker 1984; Mahadevaiah et al. 1984). Chromosome anomalies result in the interruption of male gametogenesis at pachytene or meiosis I (MI), and are therefore likely to reveal different kinds of information about events controlling the program of spermatogenesis than do autosomal genes affecting primarily spermiogenesis.

Chromosome translocations affecting spermatogenesis may be divided into two categories: the sex-chromosome anomalies, including X-autosome translocations (T[X;A]s), and the autosomal chromosome anomalies, including primarily autosome-autosome translocations (T[A;A]s) and Robertsonian fusions, but also insertions and inversions.

3.1 Sex-Chromosome Anomalies

3.1.1 Sex-Chromosome Aneuploidy

The most extreme effect of sex-chromosome aneuploidy is that caused by the presence of two X chromosomes in a phenotypic male, such as XXY (known as Klinefelter syndrome in humans) and XX,*Sxr* (sex-reversed), a condition brought about by the recurring meiotic transposition to the X of a small duplicated portion of the Y chromosome containing sex-determining genes (Cattanach et al. 1971; Eicher 1982; Singh and Jones 1982; Evans et al. 1982). Both XX,*Sxr* mice (Cattanach et al. 1971; Cattanach 1975) and XXY mice (Cattanach 1961 a; Russell and Chu 1961; but see also Slyzinski 1964; Huckins et al. 1981) have small testes completely devoid of germ cells. In the case of the XX,*Sxr* mouse, fetal testes contain abundant germ cells. These are lost in the 10 days following birth, which is the period of spermatogonial proliferation in normal XY testes (Cattanach et al. 1971; Cattanach 1975). It may be possible that the presence of two X chromosomes is incompatible with germ-cell proliferation and differentiation in a testis (Cattanach et al. 1971; Burgoyne 1978). There is no direct evidence for or against this hypothesis. One curious inconsistency in this general picture of failure of germ-cell proliferation in males with two X chromosomes is the report (Searle et al. 1983 b) of a male with the presumed karyotype $XX^{11}Y$ (unbalanced segregant from a female heterozygous for T[X;11]38H) with spermatogenic cells arresting at meiosis I and II. This may mean that the presence of two X chromosomes is not incompatible with spermatogonial proliferation and differentiation. Or it may implicate a factor on the X, distal to the T38H breakpoint, that inhibits germ-cell proliferation (de Boer 1986).

Other sex-chromosome aneuploidies that also cause male sterility are associated with either germ-cell depletion or meiotic arrest of spermatocytes. In contrast to failure in spermatogonial proliferation in XXY and XX,*Sxr* mice, the XYY mouse is characterized by degeneration of meiotic and early postmeiotic cells, and, in some cases, limited spermatogenesis (Russell and Chu 1961; Cattanach and Pollard 1969; Evans et al. 1969, 1978; Rathenberg and Muller 1973; Das and Kar 1981). In fact, one male XYY mouse was found to be transiently fertile (Evans et al. 1978). Therefore, the impact on germ cells of two Y chromosomes is not as great as that of two X chromosomes. The presence of two X chromosomes interferes with early postnatal differentiation of the testis and germ cells (spermatogonial proliferation and onset of meiosis). The presence of two Y chromosomes does not seem to interfere with this early phase of testis function, but does impede germ-cell differentiation at the time of meiosis or subsequent to meiosis. It has been suggested (Burgoyne 1979; Burgoyne and Biddle 1980) that spermatogenic failure in XYY mice results from failure in the pairing process among the three sex chromosomes and that germ cells with a univalent sex chromosome (either an X or Y) fail in subsequent differentiation. There are also other instances in which a disruption of sex-chromosome pairing has been postulated to cause spermatogenic breakdown: a mouse with failure in X-Y pairing (Beechey 1973) and fertile hybrids with a high level of X-Y dissociation (Imai et al. 1981; Matsuda et al. 1982; Matsuda et al. 1983). Normal pairing of the X and Y chromosomes may well be imperative for the completion of meiosis and the onset of spermiogenic function.

3.1.2 Sex-Chromosome Translocations

All known cases of translocations involving the sex chromosomes lead to male infertility.

There are only scant data on mice reported to have a Y-A translocation (Leonard and Deknudt 1969; Cacheiro et al. 1974; Searle 1974; Cacheiro 1983). In one case (Leonard and Deknudt 1969), all spermatogenic cells were arrested at diakinesis or MI. In another (Cacheiro et al. 1974), although the mouse was sterile, later spermatids were reported to be present. Cacheiro (1983) has reported Y-A translocations involving all of the autosomes except Chromosomes 1 and 11. All of these males were sterile, but the degree of spermatogenic impairment has not been reported. Clearly, more data are needed to clarify the effect of a Y-A translocation on spermatogenesis; however, these mice are only rarely identified and the aberration cannot be transmitted.

X-autosome translocations are rare, in both mouse and man. Their incidence is less than would be predicted on the basis of chromosome lengths and opportunity for translocation, and this is thought to be due to the fact that many such translocations are lethal due to functional hemizygosity of the autosomal portion of the translocated X chromosome in females (due to X inactivation). The presently extant murine X-A translocations are listed in Table 2 and have been reviewed extensively by Russell (1983).

All known X-A translocations act as male steriles, with a cell lethal effect prior to MII, most commonly during the pachytene stage. The only exception to this generalization is the unbalanced carriers of an insertional, nonreciprocal translo-

Table 2. Mouse X-autosome translocations

Translocation	Breakpoints[a]	Closest mapped loci X	Closest mapped loci A	References
T(X;2)14R1	XF;2C	–	pa	Cacheiro (1978); Russell (1983)
T(X;4)1R1	XF1;4A5	Ta	wi	Russell and Bangham (1961); Russell et al. (1974); Russell (1983)
T(X;4)7R1	XA2;4D1	le	m	Russell (1972); Russell et al. (1974); Russell (1983)
T(X;4)8R1	– –	le	m	Russell et al. (1974); Russell (1983)
T(X;4)37H	XA2;4D3	spf	m	Searle et al. (1983b)
T(X;7)2R1	XA4;7D3	spf	tp	Russell et al. (1974); Russell (1983)
T(X;7)3R1	XA2;7F1	spf	c	Russell and Bangham (1959); Russell et al. (1974); Russell (1983)
T(X;7)4R1[b]	?	Ta	c	Russell (1963); Russell and Montgomery (1969, 1970)
T(X;7)5R1	XF1;7A3	Ta	ru-2	Russell (1963); Russell et al. (1974); Russell (1983)
T(X;7)6R1	XF1;7B3	Ta	ru-2	Russell (1963); Russell et al. (1974); Russell (1983)
T(X;7)38H	XA2;11E1	spf	Re	Searle et al. (1983b)
T(X;12)13R1	XA3?;12A	–	–	Cacheiro et al. (1978); Russell (1983)
T(X;16)16H[c]	XD;16B5	Ta	md	Lyon et al. (1964); Eicher and Washburn (1977)
T(X;17)15R1	XA;17A	–	–	Russell (1983)
Is(7;X)1Ct[d]	XE1, XF1; 7C-7E3 or 7F1	Mo^{br}	ru-2, sh	Cattanach (1961a, 1966); Eicher and Washburn (1978)

[a] Refer to Green (1981) for cytological maps.
[b] Now extinct.
[c] Also referred to as Searle's translocation.
[d] Technically an insertion, not a translocation, but often referred to as Cattanach's translocation.

cation, Cattanach's translocation, (T[X;7]1Ct or, more appropriately, Is[In7;X]1Ct), which are fertile. The balanced carrier for this insertional translocation becomes sterile early in life (Eicher 1970; Cattanach 1974).

It is important with translocations, as with all of the genetic causes of male sterility, to know whether the deleterious effect on spermatogenesis is intrinsic or autonomous to the germ cells. This was determined by the analysis of chimeras with one cellular component genetically normal and the other carrying an X-A translocation (Russell et al. 1979, 1980b). In chimeric mice where both components were male, the component carrying the X-A translocation was never transmitted to the offspring, while the other component was. This indicated not only that the translocation is acting within the germ cells, but also that it is not producing a testicular effect and cannot be corrected by products from normal cells.

Since the deleterious effect of X-A translocations in causing male sterility is autonomous to the germ cell and spermatogenic impairment is at or around the pachytene stage, the causes of sterility should be sought in events requiring the participation of the interrupted X chromosome or of the translocated autosomal segment. However, in spite of the interruption of the X, there is no failure in pairing of the X and Y chromosome that could cause spermatogenic arrest. Quantitative analyses of pairing from microspread synaptonemal complexes was conducted for a number of X-A translocations (Ashley 1983; Ashely et al. 1982, 1983), and revealed no failure of X and Y pairing. Of particular note is the fact X-Y pairing occurs even when the translocation breakpoint on the X occurs in the Y-pairing segment of the X (Ashley et al. 1982, 1983).

Furthermore, sterility does not appear to result from inactivation of genes essential for spermatogenesis on the autosomal segments translocated to the X, which is presumed to be inactive during spermatogenesis (see Sect. 3.3). The evidence for this conclusion is provided from two sources. Russell et al. (1980a) attempted genetic "rescue" by combining each of five different X-A translocations with a Chromosome 7 tandem duplication. Since the duplication contains a significant portion of Chromosome 7, it might be presumed to substitute for possibly inactivated portions of Chromosome 7 translocated to the X. However, the males were sterile, and therefore no genetic rescue of germ-cell lethality occurred. Erickson (1984) found that Cattanach's translocation rescued the male sterility due to homozygosity for two partially-complementing albino deletions. Mice bearing these two deletions are viable but runted and sterile (Lewis et al. 1978). It is not known if the deleterious effects on spermatogenesis are intrinsic to the germ cells or a result of generally impaired vigor. The portion of Chromosome 7 which is inserted in the X chromosome in Cattanach's translocation carries the normal chromatin for the albino (C) locus. One and possibly two male mice bearing both Cattanach's translocation and the sterile albino deletions were fertile, suggesting that the translocated piece of the 7 within the X was able to "cover" for the effect of the deletions on male sterility. This inventive analysis is interesting and may indicate activity of the insertional translocation during spermatogenesis; however, there are several caveats in the interpretation of the data. First of all, most of the males tested were sterile. This could be due to the presence of the deletions, or to Cattanach's translocation, or to both. Secondly, since Cattanach's translocation and the two albino deletions are not on the same genetic background, the mice bearing both the translocation and the deletions were hybrids. It was not determined whether the deletions produced sterility on the hybrid background. (Sterility of many autosomal mutations and translocations is dependent on genetic background.) Thirdly, since it is not known if the effect of the albino deletions is intrinsic to the male germ cells, correction of sterility by Cattanach's translocation may not be due to the action (or lack of action) of the Chromosome 7 insertion in the germ cells, but may be a somatic effect.

In conclusion, thus far sterility in male carriers of X-A translocations is unexplained. It has been postulated to be due either to failure in inactivation of the X chromosome or to failure in "saturation" of pairing sites. Since both of these phenomena have also been invoked to explain sterility of autosome-autosome translocations, I shall defer a discussion of these hypotheses to Section 3.3.

3.2 Autosomal Chromosome Anomalies

Male sterility is also caused by heterozygosity for a number of autosomal chromosome rearrangements. (I shall not consider here those translocations that do not directly affect the process of spermatogenesis but do cause semisterility due to the production of chromosomally unbalanced gametes, thereby giving rise to aneuploid inviable zygotes. See reviews by Gropp et al. 1982; Chandley 1984.) Interruption of spermatogenesis as a consequence of autosome translocations was first reported by Lyon and Meredith (1966). This initial report has been followed by numerous others describing the apparently male-limited sterility of some autosome translocations in the mouse (reviews by Searle et al. 1978; Gropp et al. 1982; Chandley 1984; de Boer 1986). These sterile translocations are listed in Table 3.

Autosome rearrangements other than simple reciprocal translocations can also impair spermatogenesis. Although males heterozygous for a single Robertsonian (Rb) translocation (a centric fusion translocation) are generally fertile, those heterozygous for two or more Rb translocations (particularly with homology of one arm) show variable spermatogenic impairment (Gropp et al. 1982). The insertional autosome translocation Is(7;1)4OH impairs spermatogenesis profoundly, with germ cells dying uniformly at pachytene (Searle et al. 1983a). Tertiary trisomics originate from numerical nondisjunction in a translocation heterozygote; they have a small, extra translocation chromosome (see Green 1981). Male are generally sterile. However, one stock, $Ts(1^{13})7OH$ has been developed with fertile, though oligospermic, males (de Boer and Groen 1974; de Boer and van der Hoeven 1977). Although mice heterozygous for a single inversion are not sterile, the presence of two overlapping inversions on the same chromosome can cause disruption of spermatogenesis (Roderick 1976; Chandley 1982).

As in the case of X-A translocations, male sterility in mice bearing autosome translocations is uncomplicated by other phenotypic effects.

3.2.1 Characteristics of Male-Sterile Autosome Translocations

Not all autosome translocations give rise to male sterility; in fact, only about 20% of the autosome translocations described by Green (1981) are known to be male sterile (Table 3). T(A;A)s giving rise to male sterility share several related attributes (Searle 1982):

1. One breakpoint is located near the centromere, while the breakpoint on the second chromosome involved is fairly distal; this gives rise to long and short "marker" translocation chromosomes (see Fig. 3).
2. The MI-pairing configurations observed are a chain of four chromosomes (CIV) or a chain of three plus a univalent (CIII + I), with ring configurations less frequently observed (see Fig. 3).

Observed pairing configurations have been explained by assuming that in the smaller translocation product both segments (the segment with the centromere and the translocated segment) are so short that chiasmata are less likely to form,

Table 3. Mouse autosomal chromosome aberrations affecting male fertility

Aberration	Breakpoints	Closest mapped loci	Effects	References[a]
1. Translocations				
T(1;13)70H	1A4;13D1	1:*fz*;13:*pe*	Fertile, but reduced sperm count	de Boer (1976)
T(4;8)36H	4D2;8C1	4:*m*;8:*Os*	Most cell loss around MI, some later stages	Searle et al. (1978)
T(5;12)31H	5B;12F1	5:*Rw*;12:*Pre*-1	Reduced sperm count, some fertile	Searle et al. (1978); Forejt (1978); Beechey et al. (1980)
T(6;12)32H	6G1;12B	—	Few sperm and spermatids	Searle et al. (1978)
T(7;19)145H	7B3;19D1	7:*p*;19:*ep*	Meiotic arrest, few if any spermatids	Forejt and Gregorova (1977)
T(10;13)199H	10C1;13A1	10:*gr*;13:*bg*	Meiotic arrest, few if any spermatids	Forejt and Gregorova (1977)
T(11;19)42H	11D;19B	11:*Re*^*wa*,19:*bm*	Pachytene stage arrest	Searle et al. (1978)
T(14;15)6Ca	14E5;15B3	14:*s*;15:*uw*	Male sterility and sperm count depends on genetic background	Baranov and Dyban (1968); Forejt (1976); de Boer (1986)
T(16;17)43H	16A;17B	17:*H*-2	Meiotic arrest, very few postmeiotic cells	Searle et al. (1978); Gregorova et al. (1981)
2. Insertions				
Is(7;1)40H	1B;7B1 or B3 and 7F1	1:*fz*;7:*fr*	Pachytene arrest	Searle et al. (1978, 1983a)
3. Robertsonian translocation combinations				
a) 2 Rb's with one chromosome in common			36% are male-sterile	Evans (1976); Gropp et al. (1982)

b) Rb with reciprocal translocation heterozygote, one chromosome in common		
T(1;13)70H, +/+, Rb(11.13)4 Bnr	Sperm count reduced	Nijhoff (1981) as cited in de Boer (1986)
T(16;17)43H, +/+, Rb(16.17)7 Bnr	Male sterility of T(16;17)43H corrected	Forejt et al. (1980)
4. Tertiary trisomy and monosomy		
Ts(1^{13})70H	Fertile (but oligospermic) to sterile, depends on genetic background, MI cell death	de Boer and Branje (1979)
Ts(5^{12})31H	Sterile, few postmeiotic cells	de Boer and Branje (1979); Beechey et al. (1980)
Ms(5^{12})31H	Sterile, a few spermatids	Beechey et al. (1980)

[a] References cited are for information on male sterility, not necessarily the original report on the aberration.

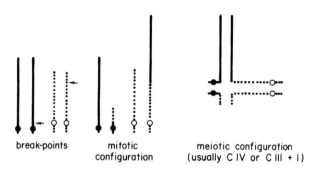

Fig. 3. The origin, and mitotic and meiotic configurations of a typical male-sterile autosome translocation. (Searle et al. 1978)

thus leading to univalence of this short chromosome at MI. Compilation of data for a number of translocation heterozygotes (much of it reported in the translocation descriptions in Green 1981) reveals that translocation heterozygotes characterized by male sterility have a higher percentage of chain configurations (CIV and CIII + I) than ring configurations and those autosome translocations associated with fertility give rise to a ring (RIV) configuration more frequently than to chain configurations.

Any hypothesis attempting to explain the basis for male sterility of various autosomal chromosome aberrations must account for a number of diverse and seemingly disparate observations which, thus far, have eluded explanation. In the following paragraphs I will describe a number of such observations that may be particularly relevant.

Sterility is found only in heterozygous carriers of certain autosome translocations; the homozygotes (when available) are not sterile. Presumably, this is so because homozygotes have normal, bivalent pairing.

In addition, the male-sterilizing effect of a translocation may not be expressed on all genetic backgrounds. A particularly notable example of this is provided by the behavior of the T(14;15)6Ca (T6) translocation. This translocation acts as a male-sterile on some genetic backgrounds and not on others (Baranov and Dyban 1968; Forejt 1976). There is evidence that a genetic factor linked to the H-2 histocompatibility complex controls this genetic variability in the expression of male sterility. Since most translocations have not been tested on different genetic backgrounds, it is not known how general this phenomenon is. Furthermore, differences in the degree of sterility caused by a translocation on different genetic backgrounds may be due to the interaction of the translocation with different genetic components. For example, de Boer and Nijhoff (1981) reported that T(1;13)7OH is fertile on one background, but sterile as a hybrid with another strain. There was a high degree of X-Y univalence noted in the translocation heterozygote hybrids (de Boer and Nijhoff 1981). Backcrosses of the hybrid translocation heterozygotes indicated that sterility and X-Y dissociation depended on the source of the Y chromosome as well as other components of the genetic background. In some cases it appears that the sterilizing effect of certain autosome aberrations can be reversed by combining them with other chromosome aberrations. When the male-sterile translocation T(16;17)43H was combined with a Robertsonian translocation involving the same two chromosomes, Rb(16.17)Bnr, the resulting double heterozygote was fertile (Forejt et al. 1979; Gregorova et al. 1981).

An additional, unexplained observation is the variability not only in the degree of azoospermia in sterile translocation heterozygotes, but also in the stage of cell death. Table 3 (see also de Boer et al. 1986) reveals that the most extreme effects are found in T42H translocation heterozygotes with arrest at the pachytene stage and in Is4OH heterozygotes where spermatocytes uniformly abort at stage IV. However, among some "sterile" translocation heterozygotes (e.g., T31H) a few fertile males can be found. The "between translocation stock" variability is compounded by a high degree of "within animal" variability, since it is frequently found that a few spermatogenic cells "escape" and develop further than the stage where most cell loss is found. Furthermore, even though the primary stage of cell death may be at or around meiotic prophase, there may be an additional effect of depletion of renewing spermatogonia (Redi et al. 1985).

Our continued inability to relate or explain these observations is reflective of a singularly poor understanding of the cause for cell death of spermatogenic cells in sterile translocation heterozygotes.

3.3 Hypotheses on Causes of Chromosome Sterility

Two theories (Lifschytz and Lindsley 1972; Miklos 1974) have been proposed to explain the sterility of both T(X:A)'s and autosome translocation heterozygotes. Neither hypothesis is wholly satisfactory.

Miklos (1974) proposed that failure in saturation of pairing sites (in the male) leads to gametocyte death. According to this theory, the existence of unpaired chromosome regions, or unsaturated pairing sites causes spermatogenic breakdown – during spermiogenesis in *Drosophila* and at an earlier stage in the mouse. Spermatogenic degeneration is limited to those cells with failure in saturating pairing sites and the extent of spermatogenic breakdown is determined by the number of unmatched pairing sites. Therefore, spermatogenic cell destiny (survival and differentiation or degeneration) is determined by a meiotic event (chromosome pairing), but may not necessarily be expressed at meiosis. The theory fails to explain specifically what "pairing sites" are, how the degree of saturation is assessed by the cell, why the effect is not as severe in females, or what is the immediate biological consequence of failure to saturate pairing sites. Many instances of chromosome sterility in the mouse seem to be explained, at least superficially, by some extension of the Miklos hypothesis. However, there are some exceptions. Chromosome insertions might be presumed to be ideal material for a test of the Miklos hypothesis, for an insertion is likely to be unpaired (i.e., to have unsaturated pairing sites) and therefore lead to gametogenic failure. In fact, one autosome insertion, Is(7;1)4OH is indeed male-sterile, with gametic arrest at the stage of pachytene. Female gametogenesis is not inhibited to nearly the same degree; however, ovarian size is smaller (Mahadevaiah et al. 1984). However, another insertion does not always lead to gametic arrest and sterility; oddly enough, this is an X chromosome insertion, Is(7;X)1Ct, commonly called Cattanach's translocation. Unbalanced carriers for this insertional translocation have two complete Chromosome 7's and an X with the insertion of a piece of Chromosome 7. According to the Miklos hypothesis, males of this constitution should be sterile since the inserted portion of Chromosome 7 lacks a pairing partner; contrary to the expectation, these males are fertile. Balanced carriers, where the inserted portion of Chromosome 7 on the X can (and sometimes does) pair with the insertion buckle on the complete Chromosome 7 (Ashley et al. 1980a, b) might be expected to be fertile; however, they are usually sterile. These observations may invalidate the Miklos hypothesis, or may simply illustrate that the unpaired chromatin of an insertion in the normally unpaired region of the X chromosome can be tolerated without deleterious effects. Another inconsistency with the Miklos hypothesis is provided by observations on the relative degree of spermatogenic impairment between mice monosomic and trisomic for the same chromosome region, $Ms(5^{12})31H$ and $Ts(5^{12})31H$ (Beechey et al. 1980). Since the extent of the chromosome region involved in the monosomy is the same as that in the trisomy, the degree of failure to "saturate" pairing should be the same in each. However, the degree of spermatogenic impairment is greater in the trisomic than in the monosomic, not predicted by the Miklos hypothesis.

Why should pairing failure lead to gametogenic cell death, and why more so in males than in females? It is commonly assumed that chromosome pairing is the means to accomplish two ends: proper disjunction of chromosomes and recombination. Failure of disjunction leads to the production of chromosomally unbalanced gametes and, subsequently, to aneuploid conceptuses. The very existence of aneuploid individuals in man, and the experimental production of mouse aneuploid fetuses from chromosomally unbalanced gametes (Gropp and Winking 1981) demonstrate that chromosomally unbalanced gametes can survive and be functional. Indeed, the ability of chromosomally unbalanced spermatozoa to fertilize had previously been demonstrated by Ford (1972). Therefore, failure in disjunction per se is not sufficient to explain the male sterility of translocation heterozygotes. This is especially true since in many cases spermatogenic disruption occurs prior to chromosome disjunction. It is also unlikely that failure in recombination in the region of translocation breakpoints is the cause of spermatogenic cell death. *Drosophila* males do not undergo recombination at all yet produce functional spermatozoa and are subject also to sterility due to translocation heterozygosity (Lifschytz and Lindsley 1972).

If we accept the hypothesis of failure in saturation of pairing sites as an explanation for translocation sterility, we must assume additional consequences of chromosome pairing other than those of assuring disjunction and recombination. Given our general lack of knowledge of meiosis it is not too unlikely that there are facets of pairing not yet understood: for example, we have only a marginal understanding of the role of DNA replication and repair synthesis during meiotic prophase and of molecular restructuring of chromatin during meiotic prophase. If these events are predicated on pairing, then failure to accomplish one or more of these transformations might act as a cell lethal.

The second theory proposed to explain chromosome sterility addresses the issue of X-chromosome inactivation during spermatogenesis. Clarification of the role of X inactivation in spermatogenesis might help in the understanding of X-autosome as well as of autosome translocation male sterility. It has long been noted that the X chromosome is heteropyknotic in spermatogenic cells, and autoradiographic studies (reviewed in Sect. 4) have documented the failure of the XY bivalent to incorporate 3H-uridine. This striking invariance in observations of X-chromosome heterochromatization and transcriptional inactivity led to the proposal (Lifschytz and Lindsley 1972) that X-chromosome inactivation is an essential feature of spermatogenesis. This hypothesis is compatible with existing data; however, it is important to bear in mind that to date the evidence for X-chromosome inactivation is not substantial, detailed, or molecular. It has not been demonstrated whether or how X inactivation is essential for the continuation of spermatogenesis. Nonetheless, this hypothesis has come to be accepted as dogma.

Failure of X inactivation has been invoked to explain the male-limited sterility of carriers of both X-autosome translocations and sterile, autosome translocations. Interruption of the X chromosome by translocations is thought to interfere with inactivation of the X chromosome occurring during meiotic prophase of spermatogenesis (Lifschytz and Lindsley 1972). In the case of male-sterile, autosome translocations, it is thought that unpaired chromosome axes might interfere with the process of X inactivation. Forejt and Gregorova (1977) have observed contact during diakinesis between the C-band of an autosome involved in a translocation multivalent and the heterochromatic C-band of the X chromosome. Similar observations of association between the extra chromosome of a tertiary trisomic male with the sex chromosomes were made by de Boer and Branje (1979); see also de Boer et al. (1986). These studies made use of both silver-stained, air-dried preparations analyzed by light microscopy and surface-spread preparations where the synaptonemal complexes were viewed by electron microscopy. When considered in the context of the Lifschytz-Lindsley hypothesis, these observations led Forejt (1982) to suggest interference with the inactivation of the X chromosome by intrusion of autosomal material into the sex vesicle. This intrusion occurs when the unpaired autosomal material associates with the unpaired end of the X chromosome. Thus, according to this hypothesis, both X-autosome translocation and autosome-autosome translocation sterility might be explained in the same manner: interference with X inactivation by intrusion of autosomal material into the sex vesicle or interruption of the X chromosome by a translocation. However, the proximity of autosomal material per se does not pose an impediment for the events occurring within the sex vesicle, for the nucleolar organizers (NORs) are always closely apposed to the surface of the sex vesicle and actively transcribing (Kierszenbaum and Tres 1974a). It has not been determined if chromosome aberrations causing sterility interfere with the relationship of the NORs to the sex vesicle, or with the transcriptive activity of the NORs.

It is not clear how the reversal of X inactivation inhibits spermatogenesis. It has been suggested (Forejt 1982) that transcription of the X chromosome leads to the accumulation of "nonpermissible" gene products, thus inhibiting the process of spermatogenesis. The use of two-dimensional gel electrophoresis analysis of proteins failed to detect the presence of additional proteins in translocation-sterile mice (Forejt 1982). However, Hotta and Chandley (1982) found elevated levels of X chromosome-coded enzymes (HPRT, G6PD, and PGK) in cell-sorted spermatocytes of sterile-translocation heterozygotes. These data on the protein products of the X chromosome are certainly suggestive of X inactivation in normal spermatocytes and a failure of that process in translocation-bearing spermatocytes. Nonetheless, detailed analysis of transcriptive activity of the X chromosome in both normal and translocation-sterile mice is needed. Speed (1986) performed autoradiography on both air-dried meiotic preparations and synaptonemal complex microspreads from sterile and subfertile males (tertiary trisomics, Ts[5^{12}]31H and Ts[1^{13}]7OH). He found evidence for abnormal transcription in the sex vesicle, which was postulated to be due either to activity of the included autosome or to reactivation of the X. Higher resolution analysis of material with greater preservation of potentially active chromatin is needed to distinguish between these alternatives and to understand the nature of the abnormal transcription observed over the sex vesicle in these sterile tertiary trisomics.

Thus, evidence is not available to validate or eliminate either the Miklos or Lifschytz-Lindsley hypotheses and the cause of translocation sterility remains unexplained. At this juncture, it should be pointed out that translocation sterility bears a formal resemblance to hybrid sterility subject to Haldane's Rule (1922). Haldane's Rule states, "When in the F_1 progeny of two animal races one sex is absent, rare or sterile, that sex is the heterozygous (i.e., heterogametic) sex." It has been demonstrated recently that the probable explanation for this type of hybrid sterility in *Drosophila* is incompatibility between the X and Y chromosome of the hybrid (Coyne 1985). Translocation sterility, like hybrid sterility, is a heterozygous sterility. It is a sterility that affects primarily the heterogametic sex. It is a sterility where the nature of the interaction between the X and Y chromosomes is implicated. Is it possible that in species hybrids there is failure in recognition of the pairing segments of the X and Y chromosomes, hence a failure in the inactivation of the X? We need more documentation of pairing failure and the transcriptional status of the X chromosome in both sterile-species hybrids and translocation-sterile mice. However, far more important, we need to understand how these phenomena might have an impact on the progress of spermatogenesis and why translocations act as meiotic cell lethals. Clearly, there is yet much to learn about both hybrid and translocation sterility, the process of meiosis, and its role in the developmental program of spermatogenesis.

4 Structure and Role of the Sex Chromosomes in Spermatogenesis

Consideration of the structure and function of the sex chromosomes during spermatogenesis is a logical sequitur to the preceding sections on chromosome sterility and the involvement of the sex chromosomes in the etiology of sterility. In this section I shall discuss the structure and significance of the sex vesicle, review the evidence for allocyclic behavior of the sex chromosomes during meiotic prophase, consider again the issue of X-chromosome inactivity during spermatogenesis, and, finally, raise the question of whether the Y chromosome has a functional role in spermatogenesis.

4.1 The Sex Vesicle

One cannot discuss the structure and function of the sex chromosomes during mammalian spermatogenesis without considering the sex vesicle, or that region of nucleoplasm within which the sex chromosomes are sequestered during much of meiotic prophase. The term "sex vesicle," introduced by Sachs (1954) has persisted in spite of Solari's (1974) admonishment that it does not accurately describe a body which is not membrane-bound. Although Solari's term "XY body" is more accurate, I shall adopt the more frequently used term, "sex vesicle."

The sex vesicle, and the condensed sex chromosomes, are not apparent until the zygotene stage of meiotic prophase, when maximal synapsis of the autosomes occurs. During the zygotene phase, the cores of the X and Y chromosomes are clearly visible within the sex vesicle, with the common end exhibiting a synaptonemal complex. Both the paired and free ends of the X and Y chromosomes are attached to the nuclear envelope. By early pachytene, the synaptonemal complex between the sex chromosomes is clearly visible and the chromatin of the X chromosome can be distinguished from that of the Y chromosome (Solari 1970). At the midpachytene stage, the common segment is very short and the core of the X chromosome is convoluted, with two components separable in an anomalous synaptonemal complex-like configuration (Solari 1970). It is at this point during pachytene that the nucleolar region is becoming highly differentiated, with a fibrillar and granular region, and is closely apposed to the inner aspect of the sex vesicle (Fig. 4). By late pachytene, the nucleolar region is well developed, and elements of it intrude into the sex vesicle. During the diplotene stage, the X and Y chromosomes become strongly condensed (Solari 1974) and by diakinesis the two chromatids of the X and Y chromosomes are visible.

Little is known of the biochemistry of the sex vesicle. It is, of course, Feulgen-positive (Sachs 1954; Solari and Tres 1967), and staining with acridine orange reveals differences in the pattern of chromatin organization along the length of the X and Y chromosomes (Solari and Tres 1967). Histochemical tests and autoradiography after exposure to tritiated uridine both failed to reveal RNA in the sex vesicle, although the closely adjacent nucleolus is positive for RNA. Tres (cited in Solari 1974) obtained evidence for the localization of basic proteins within the sex vesicle. It is not clear whether there is a special substance of the sex vesicle, or whether, as is more likely, the sex vesicle structure is simply a reflection of the state of the chromatin of the X and Y chromosomes, which clearly manifest allocycly to the autosomal chromosomes, both with respect to condensation and meiotic behavior. In male mice bearing the X-autosome translocation T(X;16)16H, the autosomal portion attached to the X becomes positively heteropyknotic, assuming a chromatin structure like that of the sex chromosomes, although it lies outside of the sex vesicle (Solari 1971). This observation suggests that the structure of the sex vesicle is a reflection of, rather than a determinant of, sex chromosome structure.

Related to the question "what is the sex vesicle?" is the question "what determines the formation of the sex vesicle?" Until recently, it has been assumed that the presence of a Y chromosome in a germ cell entering meiosis determined the formation of a sex vesicle. That this is not the case has been shown by Hogg and

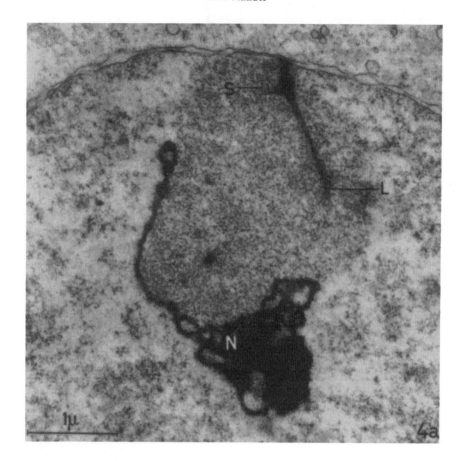

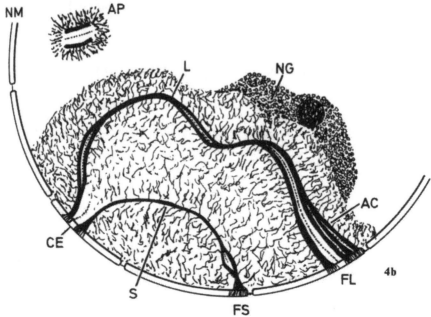

McLaren (1985), who examined germ cells ectopically located in the fetal adrenal glands of mice. These germ cells, presumably under the influence of adrenal hormones, enter meiosis prenatally (on an "oogenesis schedule") in both male and female fetuses. Prenatal meiotic germ cells of XY male fetuses were examined and found not to contain a sex vesicle, in spite of the presence of a Y chromosome. Furthermore, Liming and Pathak (1981), in a study of sterile, interspecific hybrids between the Indian and Chinese muntjacs, found that the hybrid males lacked a sex vesicle in spite of the presence of a Y chromosome. Sex vesicles are present in the spermatocytes of XO,*Sxr* mice, possessing only a very small portion of the Y chromosome (Kot and Handel, unpublished). These observations suggest that the Y does not determine and may not even be required for sex vesicle formation. Therefore, it appears that the sex vesicle is a structure unique to spermatogenesis, as opposed to oogenesis. Furthermore, the presence of a sex vesicle also does not depend on the formation of a synaptonemal complex between the sex chromosomes, for a sex vesicle is present in at least two species exhibiting no synapsis between the sex chromosomes, the sand rat (Solari and Ashley 1977) and the southern pygmy mouse (Pathak et al. 1980). It, therefore, appears likely that the sex vesicle is a reflection of a unique state of the chromatin of the sex chromosomes during spermatogenesis.

4.2 Allocyclic Behavior of the Sex Chromosomes

The pattern of condensation, transcription, and behavior of the sex chromosomes during spermatogenic meiosis is in striking contrast to that of the autosomal chromosomes. This allocycly is noted as early as the premeiotic S phase (prior to sex vesicle formation). During the premeiotic S phase, the X replicates later than the autosomes, there is an even later replicating segment of the X adjacent to the centromere, and the Y chromosome begins and ends replication later than the X chromosome (Kofman-Alfaro and Chandley 1970). The significance of late replication is still unclear, but it is a property of heterochromatin.

At the onset of the pachytene stage of meiosis, the sex chromosomes are condensed and heteropyknotic in comparison to the autosomes. This condensation is correlated with their transcriptive inactivity. Autoradiography at the level of both light microscopy (see Monesi et al. 1978 for review) and electron microscopy of sectioned and microspread pachytene cells (Kierszenbaum and Tres 1974a, b) has demonstrated that neither the X nor the Y chromosomes incorporate 3H-uridine during the pachytene stage of meiotic prophase.

Fig. 4a. Electron micrograph of a late pachytene stage mouse spermatocyte showing the arrangement of chromosome axes of the sex vesicle. *L* long axis (the X chromosome); *S* short axis (the Y chromosome); *N* nucleolus (Solari 1970). **b** The sex vesicle during the midpachytene stage. *L* long axis (the X chromosome); *S* short axis (the Y chromosome); *CE* common (paired) end with a short synaptonemal complex; *FL* and *FS* free ends of the long and short axes, respectively; *AC* anomalous synaptonemal complex of the long axis; *NG* nucleolar granules; *AP* autosome pair; *NM* nuclear envelope membrane (Solari 1970)

4.3 X-Chromosome Inactivation as a Correlate of Spermatogenesis

Since in meiotic cells the X chromosome is heteropyknotic, replicates late, and fails to incorporate 3H-uridine, it is assumed to be inactive. In Section 3.3 I presented a theory (Lifschytz and Lindsley 1972) stating that X-chromosome inactivation is an essential correlate of gametogenesis in the heterogametic sex. I have pointed out that although the supporting evidence is not substantial, it is not contradictory and the theory has come to be accepted as dogma. It may well be correct; however, more data are needed. I will now critically consider several aspects of this theory and the evidence from mouse spermatogenesis.

The molecular basis of X-chromosome inactivation is not understood either for inactivation of X chromosomes in the somatic cells of mammalian females or for the inactivation of the X chromosome in spermatogenic cells; available evidence suggests that the two forms of inactivation are not the same. Although the inactive X chromosome of female somatic cells is unable to transform cells, the DNA from the X chromosome of spermatogenic cells and mature sperm is able to function in DNA-mediated transformation of hypoxanthine-guanine phosphoribosyl transferase (HPRT) negative cells (Venolia et al. 1984). These data suggest that whatever the chromatin modification that causes inactivation (DNA methylation and/or chromosomal proteins), the modification is not as stable in the spermatogenic, inactive X as it is in the female somatic cell, inactive X.

Not only is the molecular basis of X-chromosome inactivation during spermatogenesis unknown, the precise time of inactivation is also unknown. It has been generally assumed, although in the absence of specific data, that the time of inactivation is meiotic prophase, when the X is observed not to incorporate 3H-uridine. However, the transcriptive activity of the X chromosome prior to this time (e.g., in preleptotene spermatocytes or in spermatogonia) has not been determined. The fact that the X is late-replicating during the preleptotene S phase suggests that it may have been inactivated at or prior to that time. The failure of prespermatogonia with two X chromosomes to proliferate and differentiate as definitive spermatogonia (see Sect. 3.1) suggests the possibility that X inactivation might be an early postnatal event associated with the proliferation of prespermatogonia (and that it does not occur in cells with two X chromosomes).

I considered earlier the suggestion that the X chromosome might be reactivated (or fail to be inactivated) in translocation-sterile mice and that reactivation acted as a cell lethal. It is not at all clear why the X should be inactivated during spermatogenesis in the first place and why its reactivation would act as a cell lethal. Expression of the X chromosome is essential for all other cells. Is it possible that transcripts and protein products of the X chromosome are indeed "nonpermissible" in the context of spermatogenesis? On the contrary, since the X chromosome codes for many "housekeeping" proteins, there seems to be no reason why expression of the X might be any less essential for survival of spermatogenic cells than for somatic cells. In fact, it appears that perhaps the sperm cell needs alternative sources of information for these housekeeping proteins in order to cope with the inactivation of the X chromosome. For instance, there is a sperm-specific variant of phosphoglycerate kinase, PGK-2, that is autosomally coded and activated only in haploid germ cells (Van de Berg et al. 1976; Eicher et al.

1978; Kramer 1981; Kramer and Erickson 1981). This evidence suggests that it is not the protein products of the X chromosome that are inimical to spermatogenesis.

To summarize, it is not apparent why the X chromosome should be inactive in spermatogenic cells, it is not known how it is inactivated, and it is not known when it is inactivated. It may be that transcriptional inactivation of the X chromosome in spermatogenic cells is simply a correlate of some other event essential for spermatogenesis. It is interesting to contrast the transcriptional inactivity of the X and Y chromosomes with the widespread, seemingly almost global, transcriptive activity of the autosomes. Similar ubiquitous transcription is also a feature of meiotic prophase of oogenesis. Is it possible that this transcription serves a function other than providing templates? And that in spermatogenic cells this does not and cannot occur on the X chromosome? In order to solve the enigma of apparent X-chromosome inactivation in spermatogenic cells, it would be useful to know if X inactivation is an event essential for spermatogenesis itself, or if it is an event that is important in preparing the paternal half of the zygotic genome. One essential event of spermatogenesis is the paternal imprinting of the entire genome, including the X chromosome (since the behavior in the zygote of the paternal X differs from that of the maternal X). It is now well documented (McGrath and Solter 1984; Surani et al. 1984) that mammalian embryogenesis cannot occur without the participation of both a male and female genome. The haploid genomes remain imprinted as maternal or paternal for a considerable period of embryonic development (Cattanach and Kirk 1985; Surani et al. 1986) and perhaps retain their original imprinting until it is removed and replaced at the time of gametogenesis. It is possible that genome imprinting is a meiotic event; and, if so, the heteropyknotic state of the X chromosome might be of relevance with respect to this process. Lack of transcriptive activity of the spermatocyte X may permit a unique imprinting (perhaps reflected in the maintenance of the transforming activity of X chromosomal DNA) that identifies the chromosome as the paternal X chromosome.

4.4 Role of the Y Chromosome in Spermatogenesis

A critical, unanswered question is "does the mammalian Y chromosome exert any genetic control over spermatogenesis?" In contrast to the Y of *Drosophila*, to which a number of male fertility factors map (for review, see Hennig 1985 and Hackstein, this volume), few genes map to the mammalian Y and no known spermatogenic factors are on the Y. In fact, the only genes known to be on the mouse Y chromosome are a gene controlling the H-Y surface antigen, the testis-determining gene, and perhaps, the Y-linked allele of the steroid sulfatase gene (Keitges et al. 1985). In humans, MIC2Y, the structural gene for a cell surface antigen, 12E7, and several RFLPs have been mapped to the Y (Bishop et al. 1985). The gene MIC2Y is the only structural gene unambiguously assigned to a mammalian Y chromosome. It is not my intention to review here the genetic structure of the mouse Y chromosome, but rather to consider the issue of whether the Y chromosome has a genetic function in spermatogenesis.

There is interesting evidence implicating a function of the Y chromosome in sperm-fertilizing ability (Krzanowska 1972, 1986). However, evidence for the presence of specific spermatogenesis factors, or sperm morphogenesis and/or motility factors on the Y chromosome, comes from studies of XO males lacking portions of the Y chromosome. The first is that produced by the *Sxr* "gene" (actually a piece of rearranged Y chromosome; see Sect. 3.1). Testes of XO,*Sxr* mice are characterized by a limited extent of spermatogenesis and the production of morphologically abnormal, nonmotile spermatozoa (Cattanach et al. 1971). Another abnormal Y chromosome, designated Y* (Eicher 1982, Eicher et al. 1983), arose by an intrachromosome rearrangement in which the distal end, observed normally to pair with the X, was moved to a more proximal location, adjacent to the testis-determining, *Tdy*, gene. Upon recombination with the X in XY* males, two recombination chromosomes are produced, X^Y and Y^X. Mice inheriting the X^Y chromosome develop testes due to the presence of the *Tdy* gene in the portion of the Y attached to the X. Whereas XX^Y males are sterile (presumably due to the presence of two X chromosomes), some $X^Y O$ mice are fertile and produce motile sperm. This production of motile sperm, in contrast to the failure of XO,*Sxr* mice to produce motile sperm, has been offered as evidence that there is a gene (or genes) on the Y chromosome controlling sperm motility. However, the issue of the numbers of surviving and differentiating germ cells in these two XO testes must also to be considered. All mice bearing mutations or chromosome aberrations leading to limited sperm production are characterized by high levels of morphologically abnormal and nonmotile sperm, such as observed in XO,*Sxr* males. In some cases this may be a gene-specific, germ-cell autonomous effect; in others, it may simply be a reflection of low numbers of differentiating germ cells and comcomitant deficiency in intraseminiferous tubule feedback and cell signaling. Seminiferous tubules are highly organized structures and adequate spermatogenesis is functionally dependent on intimate contact and relationships of the germ cells and Sertoli cells. There is considerable evidence that Sertoli cell secretions are functions of the stage of the seminiferous epithelium cycle, i.e., the stage of spermatogenic cell development (LaCroix et al. 1981; Ritzen et al. 1982; Wright et al. 1983; Griswold 1987). Since it is probable that signals transmitted from germ cells to Sertoli cells result in Sertoli cell secretion of factors creating the appropriate microenvironment for each stage of germ-cell development, inadequate numbers of maturing germ cells may well result in a failure of the Sertoli cells to provide the correct microenvironment for the germ cells. Such may possibly be an alternative explanation for the failure of germ cells in an XO,*Sxr* testis to develop normally and acquire motility.

The Y chromosome has also been implicated in the control of spermatogonial proliferation. A variant of *Sxr* mice, known as *Sxr'*, lack the genetic determinant for the H-Y antigen on the *Sxr* piece of the Y chromosome (McLaren et al. 1984). XO,*Sxr'* mice, in contradistinction to XO,*Sxr* mice, are deficient in replicating spermatogonia and in cells entering meiosis (Burgoyne et al. 1986). This had led to the suggestion that the H-Y gene may in fact be a "spermatogenesis" gene (Burgoyne et al. 1986). However, since the genetic extent and complexity of neither the *Sxr* nor the *Sxr'* piece of the Y chromosome are known, the missing "factor" may be a gene adjacent to H-Y rather than the H-Y gene itself. Nonetheless, the fact that the deletion of a small piece of DNA leads to failure in spermatogonial replication is a significant and intriguing observation. It is possible that this is indeed an elusive genetic "spermatogenesis factor"; but in this instance, what does that mean? Is it a factor acting autonomously in the male germ cell, controlling early proliferation of the germ line? Or is it a gene controlling the activity of the X chromosome (a distinct possibility, since the phenotype of the early postnatal XO,*Sxr'* testis is similar to that of testes with two X chromosomes)? Or is it a gene not acting autonomously within the germ cell, but controlling the synthesis and secretion of a Sertoli cell mitogenic factor (such as that described by Feig et al. 1980)?

In summary, there is only limited but not yet definitive evidence for the presence of genes on the Y chromosome controlling spermatogonial proliferation and spermiogenic differentiation. Identification of specific spermatogenesis genes on the Y chromosome requires more genetic data and specific information about the differentiation of spermatogenic cells in mice lacking portions of the Y chromosome.

5 Overview of Molecular Events of Spermatogenesis

In the preceding sections I have covered genetic conditions – both autosomal mutations and chromosome aberrations – affecting spermatogenesis and have discussed the evidence from these conditions for genetic control of the differentiative processes of spermatogenesis. I will now turn to the biochemical events of spermatogenesis for evidence of direct genetic control. It is clear that an evolving understanding of the genetic regulation of spermatogenesis will rely on both genetic and molecular evidence.

In this section I will briefly review the molecular events of spermatogenesis, with an emphasis on the synthetic activities during meiosis. In the next section I will consider the evidence for haploid action of genes during spermiogenesis.

5.1 Spermatogonial Stages

The spermatogonia from a continually renewing cell population that also give rise to cells committed to undergo meiosis (type B spermatogonia). Determining the kinetics of cell renewal and the trigger for the onset of the premeiotic path of development has been and continues to be a major biological problem. Furthermore, virtually nothing is known of the molecular biology of spermatogonial growth and differentiation. This is not a trivial gap in our knowledge, for it may well be that a determinative event for spermatogenesis occurs during spermatogonial cell renewal and differentiation. Insights into this problem may be provided by continued analysis of various genetic models, such as male mice with two X chromosomes and the XO,*Sxr'* mice that lack part of the *Sxr* piece of the Y chromosome and fail in proliferation of prespermatogonia.

5.2 Spermatogenic Meiosis

The last division of type B spermatogonia yields preleptotene spermatocytes that undergo DNA replication (the preleptotene S phase) and then enter meiotic prophase. Prophase of meiosis is lengthy, 11 to 12 days in the mouse, and characterized by intense RNA transcription activity. Prophase is followed relatively rapid by the remaining meiotic stages, resulting in the formation of four haploid spermatids.

5.2.1 DNA Synthesis During Meiosis

The premeiotic S phase is qualitatively and quantitatively different from mitotic S phases: it is more protracted and at its termination there remain unreplicated segments of the genome whose replication is not completed until meiotic prophase. Studies of meiosis in the lily have revealed two fractions of DNA with distinctive replication patterns during meiotic prophase. The first of these is the "zygotene DNA," or zygDNA, (about 0.1–0.2% of the genome) that undergoes semiconservative replication during the period of chromosome pairing (Hotta et al. 1966). The second subset of DNA is that which undergoes repair replication during the pachytene stage, and may therefore be involved with recombination. Similar subsets of DNA have now been identified in mouse spermatocytes (Hotta et al. 1985; Stubbs and Stern 1986).

DNA with similar S1 nuclease sensitivity as lily zygDNA has been identified in isolated mouse spermatocytes. Furthermore, mouse spermatocytes have a fraction of poly(A)+RNA which shows similarities to the unique lily zygRNA that is transcribed from zygDNA (Hotta et al. 1985). The mouse zygRNA is not detected in premeiotic spermatocytes or in postmeiotic spermatids (Hotta et al. 1985). The function of zygRNA has not been clearly established for either lily or mouse. It is hypothesized, primarily on the concordance of its synthesis with chromosome pairing, to function in the pairing process, possibly as a means of chromosome recognition permitting effective pairing. Interestingly, this fraction of RNA is reduced in an achiasmatic lily hybrid (Hotta et al. 1985).

DNA undergoing repair replication in mouse pachytene spermatocytes was characterized (Stubbs and Stern 1986). It was shown that these DNA repair sites are a selective subset of the mouse genome, and have a distinct chromatin structure, evidenced by DNase II sensitivity and Mg^{2+} solubility of the chromatin. Again, these characteristics are similar to those exhibited by the DNA repair sites in the lily, suggesting extraordinary evolutionary conservation of molecular events of meiosis.

5.2.2 RNA Transcription During Meiosis

Much of our information on the transcriptive activities of mouse spermatocytes stems from the careful autoradiographic observations of Monesi and coworkers (reviewed by Monesi et al. 1978). A drop in RNA transcription occurs after the preleptotene S phase, followed by a rapid rise in the incorporation of 3H-uridine to a peak at the midpachytene stage and another decline in activity during diplotene and diakinesis. Little or no synthesis of RNA occurs during metaphase and anaphase. These results have been confirmed by elegant, high resolution, autoradiographic studies of Kierszenbaum and Tres (1974b), demonstrating that RNA synthesis occurs in a periaxial site on chromatin loops emanating from the axes. The RNA synthesized during the pachytene stage is remarkably stable, remaining associated with the chromosomes for a long period of time and then rapidly transported to the cytoplasm at diakinesis or prometaphase (Monesi et al. 1978). By examining RNA profiles of separated and semipurified spermatogenic cell populations after labeling with 3H-uridine in vivo or in vitro, Monesi showed that the labeled RNA was heterodisperse, including polyadenylated RNA.

Little is known about specific mRNA sequences transcribed during meiotic prophase. This information may well be pivotal for an understanding of how the events of spermatogenesis are regulated, since, as previously pointed out, many of the RNA sequences transcribed during the spermatocyte phase are retained during subsequent spermiogenesis. The spermatid-specific PGK-2 is synthesized during or prior to meiotic prophase, since the mRNA can be immunoprecipitated from a nonpolysomal fraction obtained from spermatocytes (Gold et al. 1983a). Likewise synthesized at this time is the transcript for the sperm-specific LDH-C_4 (Wieben 1981). It is also assumed that the messenger RNAs for histones (Bhatnagar et al. 1985), testicular cytochrome c (Goldberg et al. 1977) and actins (Hecht et al. 1984), are synthesized at or prior to the pachytene stage since their protein products can be detected in pachytene spermatocytes.

Mice bearing chromosome aberrations causing male-limited sterility, where spermatogenic breakdown frequently occurs during meiotic prophase, may be useful models for understanding the importance of molecular events of meiotic prophase. For instance, it is known that there are irregularities of pachytene stage DNA metabolism in these mice, with excessive and random nicking of DNA (Hotta et al. 1979). Activities of certain X-linked enzymes may be elevated in such mice (Hotta and Chandley 1982), suggesting abnormalities of RNA transcription. However, virtually nothing is known about general patterns of RNA transcription in mutant mice in which spermatogenesis is disrupted at the spermatocyte stage.

5.3 Spermiogenesis

During postmeiotic differentiation of spermatogenic cells, an impressive and intricate program of cytodifferentiation unfolds, resulting in the streamlined spermatozoon, with a condensed nucleus enfolded by the acrosome and the motile sperm tail with its ordered array of microtubules and accessory fibers. Later, I will consider in detail an important unanswered question, namely, how much of this cytodifferentiation might be dependent on concurrent gene transcription (haploid gene action) and how much is due to expression of stored templates and proteins, presumably synthesized during meiosis?

5.3.1 RNA Synthesis During Spermiogenesis

Extensive autoradiographic analyses from Monesi's laboratory have shown that RNA synthesis, halted in late meiotic prophase, resumes in early round spermatids (steps 1–8) and then completely halts in elongating and condensing spermatids (steps 9–16) (evidence reviewed by Monesi et al. 1978). The level of RNA synthesis in round spermatids, expressed relative to DNA content, is approximately equal to that of late pachytene spermatocytes. For both classes of cells, the newly synthesized RNA includes rRNA and polyadenylated as well as non-polyadenylated RNAs; since the latter RNAs are associated with polysomes, they presumably include mRNAs (Monesi et al. 1978; D'Agostino et al. 1978). Gold et al. (1983b) extracted poly(A)+RNA from a variety of spermatogenic cells purified by unit gravity sedimentation and Percoll gradient centrifugation. Polypeptides synthesized from these RNAs in a cell-free translation system were displayed by two-dimensional gel electrophoresis and comparisons between pachytene spermatocytes and round spermatids revealed that a minimum of 5–10% of the polypeptides are either specific to or highly enriched in each cell type.

The autosomal recessive mutations discussed in Section 1 all cause spermiogenic abnormalities that are manifest cytologically. When gene-specific probes are available, the analysis of mice bearing these mutations may be a promising avenue toward understanding the importance and role of various genes in the cytodifferentiation of spermatids. For instance, although spermatids of homozygous *qk* mice are characterized by abnormal head morphology, the synthesis of mouse protamine occurs in the testes of these mutants (Hecht et al. 1985). Therefore, abnormal nuclear shaping is not due to failure in the synthesis and processing of mouse protamine 1. Causes of abnormal head shape must be sought in the organization of chromatin or elsewhere.

6 Haploid Gene Action During Spermatogenesis

In this section I shall review the evidence for haploid gene action during spermatogenesis and consider the implications for sperm phenotype and genetic transmission. It is important in this context to distinguish between those genes that may exhibit haploid gene expression (i.e., the gene product is first manufactured during the haploid, or postmeiotic phase of sperm differentiation) and those genes that exhibit haploid gene action (i.e., the gene is first transcribed during the haploid phase). In the first instance, since messengers are transcribed prior to meiotic divisions, meiotic products are not different from one another by virtue of the products of these genes. Detection of quantitative and qualitative differences in proteins between meiotic and postmeiotic cells, with some proteins present only in postmeiotic cells (Boitani et al. 1980; Kramer and Erickson 1982) is evidence for haploid gene expression. However, it may not constitute evidence for haploid gene action if the proteins synthesized during spermiogenesis are translated from stored messengers. When a gene is transcribed postmeiotically (haploid gene action), the meiotic products can be different, reflecting different alleles on the segregated chromosomes. Synthesis of mRNA postmeiotically constitutes evidence for haploid gene action. Some products which appear to be transcribed during spermiogenesis are also transcribed during meiosis (e.g., LDH-C$_4$). Haploid action of specific genes has been detected by both genetic analysis and direct biochemical analyses using specific gene probes.

6.1 Postmeiotic Synthesis of RNA

As reviewed previously (Sect. 4.3) RNA synthesis declines after meiosis and is not detected in late-stage spermatids. Studies utilizing separated testicular cells labeled in vivo (Erickson et al. 1980) suggest that, normalized to DNA content, the rate of RNA synthesis in early round spermatids is equal to that during the pachytene stage. Functional assays suggest both the expression and synthesis of new mRNAs during this phase of differentiation. Fujimoto and Erickson (1982) enriched pachytene spermatocytes and round spermatids by centrifugal elutriation and analyzed cell-free protein synthesis products from RNAs purified from these cells by two-dimensional gel electrophoresis. Analysis of the gels revealed qualitatively new proteins in the round spermatids, suggesting at least the expression and possibly the synthesis of new mRNAs by these cells. In similar studies, Stern et al. (1983 a, b) compared polypeptides synthesized in vivo and in vitro from both round spermatids and elongating spermatids and found that approximately 5–10% of the synthesized polypeptides differed quantitatively between the two cell types. These results suggested modulation of gene expression by translational regulation and differential utilization of stored mRNAs after the cessation of RNA synthesis during spermiogenesis.

6.2 Molecular Evidence for Transcription of Genes During the Haploid Phase

Firm evidence for haploid gene transcription depends on the demonstration of newly synthesized and functional mRNA species present in spermatids and not present in meiotic spermatocytes. The use of specific cDNA probes to characterize extracted RNA has provided evidence for the presence of mRNAs present in spermiogenic cells, but not detectable in spermatocytes. It is anticipated that sensitive techniques of in situ hybridization (Gizang-Ginsberg and Wolgemuth 1985) will provide additional information. However, evidence about the time of synthesis and utilization of these templates is generally not available. This kind of evidence is critical in establishing haploid gene action.

A differential screen of a testis cDNA library was conducted by Kleene et al. (1983). The testis cDNA library was prepared by synthesis of cDNAs from total testicular cytoplasmic poly(A)+ RNA. Pachytene spermatocytes and round spermatids were isolated (using sedimentation at unit gravity followed by density gradient sedimentation on Percoll gradients) and used as a source of poly(A+) RNA for cDNA synthesis. These cell-specific cDNAs were used to screen the testicular library for the detection of clones reacting with round spermatids and not pachytene spermatocytes and vice versa. In this screen, all clones reacting with pachytene-derived cDNAs also reacted with round spermatid-derived cDNAs, thus failing to provide evidence for pachytene-specific transcripts. However, a number of clones reacted with round spermatid cDNA about tenfold more strongly than with pachytene cDNA, thus representing cDNA clones of transcripts either unique to or enriched in cells in the haploid phase of differentiation. A similar strategy was used by Dudley et al. (1984) to identify haploid stage mRNAs. A testis cDNA library was screened with liver and testis RNA to detect clones for RNA expressed ten fold or greater in the testis than in the liver. These clones were then used to screen RNA from the testes of 2-week-old and 3-week-old mice in order to detect those sequences expressed predominantly during the spermiogenic phase of differentiation. This was further screened using RNA from the testes of germ-cell deficient XX,Sxr and Tfm/Y mice.

6.2.1 Specific Genes Transcribed During the Haploid Phase

Use of cDNA or genomic clones for known genes to probe mRNA of spermatogenic cells can provide evidence for the transcription of specific genes during the haploid phase of differentiation.

Tubulin. Distel et al. (1984) isolated a complementary DNA clone for α-tubulin from a mouse testis cDNA library using a clone containing the coding sequences and the 3' untranslated sequence from rat brain α-tubulin. This mouse testis cDNA clone was homologous to much of the rat brain probe; however, the 3' untranslated region did not hybridize to the rat probe. Since other experiments have demonstrated that the 3' untranslated sequences of tubulins may reveal specific transcripts (see Distel et al. 1984) this region was subcloned and used as a probe for mRNA species produced during spermatogenic differentiation. The probe hybridized specifically to testicular RNA and not RNA from brain. Hybridization occurred with two transcripts, 2.1 kb and 1.55 kb in length, that were enriched in round spermatids in comparison to pachytene spermatocytes. These unique transcripts may encode an α-tubulin utilized in either the spermatid manchette or the axoneme.

Actin. Hecht's laboratory also has evidence for haploid accumulation of the message for a unique isoform of actin (Waters et al. 1985). A probe specific for mouse γ-actin detected a 2.1-kb mRNA present in testes of all ages (Hecht et al. 1986) and in isolated germ cells of different stages (Waters et al. 1985) as well as a 1.5-kb transcript present only in postmeiotic spermatids. The 2.1-kb mRNA encodes both β- and γ-actin. The 1.5-kb transcript, unique to haploid cells, was not detected by specific isotype probes of the 3' untranslated region of β- and γ-actin. It may be a modified form of either the β- or γ-actins detected in mouse spermatogenic cells (Hecht et al. 1984) or may represent further processing of the 2.1-kb mRNA. It is also possible that there is a unique haploid actin that plays a role in the formation of the sperm tail, acrosome, or postacrosome region, where actin has been localized by immunofluorescence in mammalian spermatids (Clarke and Yanigimachi 1978).

Protamines. Two of the haploid-specific cDNA clones identified by Kleene et al. (1983) were sequenced and demonstrated to be coding for mouse protamine 1 and 2 (Kleene et al. 1985; Yelick et al. 1985). It was shown that protamine 1 RNA is unique to or enriched in haploid cells and that it is subject to translational regulation. It is presumed to be stored as an untranslated ribonucleic protein in round spermatids and then partially deadenylated prior to or concurrent with translation in elongating spermatids (Kleene et al. 1984).

PGK-2. The mRNA for sperm-specific PGK-2 has been detected using precipitation by an antibody specific to this isozyme (Gold et al. 1983a). These data indicated that the messenger for this isozyme first appears in meiotic spermatocytes and increases in amount during the haploid phase. Therefore, transcription is not unique to the haploid phase, but the amount of the transcript increases during this phase. Since the message is not translated until the haploid phase of differentiation (Kramer 1981; Kramer and Erickson 1981), this message also appears to be subject to translational regulation.

Transcription of Other Genes. Fujimoto et al. (1984) analyzed the expression of RNA complementary to a clone, pM459, obtained by inserting homopolymeric (dT)-tailed mRNA-cDNA hybrid molecules into pBR322. The source of mRNA was poly(A+) RNA prepared from isolated spermatids more than 90% pure. Cytosolic dot blots gave evidence that the RNA complementary to this clone is enriched in testes from 21-day-old mice as opposed to 14-day and 7-day testes. Labeled RNA complementary to the pM459l clone is present in spermatids which have been incubated in vitro with 3H-uridine. Clone pM459 has recently been sequenced (Tanaka and Fujimoto 1986) and the obtained sequence closely matches the carboxy-terminal residues of mouse LDH-C_4.

Hecht et al. (1986) identified in sexually mature testes an abundant RNA recognized by one of the haploid-specific cDNAs detected in their 1983 study (Kleene et al. 1983). This as yet uncharacterized cDNA, called 11A, hybridizes to a 620-base mRNA first detected in testes during the first wave of spermiogenesis.

Using a probe for an oncogene, *c-abl,* Ponzetto and Wolgemuth (1985) detected a unique transcript in postmeiotic spermatogenic cells. The transcription of this gene is not specific to the testis, or to postmeiotic germ cells. However, the size of this particular transcript is specific to spermatids.

Another gene product abundant in haploid germ cells is the *t* complex polypeptide 1 (TCP-1), a nonglycosylated external membrane protein (Silver and White 1982). The gene has recently been cloned and sequenced (Willison et al. 1986); however, analysis of the time of transcription is not available. Since the transcripts of this protein are abundant in haploid cells, but not confined to these cells (or even to the testis), this may be an example of a gene exhibiting haploid expression but not necessarily haploid transcription.

6.3 How Significant is Haploid Gene Action During Spermatogenesis?

In considering the significance of haploid gene action, specifically in the creation of two or more populations of spermatids, several issues must be evaluated. These include the presence of cytoplasmic bridges connecting spermatogenic cells, whether or not the action of a gene is restricted to the haploid phase, and evidence from genetic analysis.

6.3.1 Cytoplasmic Bridges

It has been demonstrated (Dym and Fawcett 1971) that at some point during the series of mitotic divisions of spermatogonia, cytokinesis ceases to be complete and all of the ensuing cells are joined by cytoplasmic bridges (Fig. 5). There has been no determination of the exact number of spermatids connected by these cytoplasmic bridges, although it is estimated that the number may be 128 or higher.

The presence of cytoplasmic bridges has frequently been invoked as a mechanism to preserve gametic equality in spite of haploid gene action (Erickson 1973). This can certainly be so in many if not most cases. Many proteins are freely soluble and diffusible in the cytoplasm and although produced in one cell, could easily be translocated via the cytoplasmic bridges to other cells. For such diffusible gene products, developing spermatids could be viewed as a functional syncytium, with little or no differentiation between component cells. However, if a protein product of a gene is packaged in an intracellular compartment such that its mobility is limited (e.g., by being inserted into the plasma membrane or associated with the cytoskeleton) then the haploid gene product of one meiotic segregant would not diffuse through a bridge to other meiotic segregants. For such gene products, interconnected spermatids are not a functional syncytium, but are a group of distinct cells, with the possibility of being different among themselves by virtue of the haploid action of genes specifying nondiffusible protein products. Therefore, although cytoplasmic bridges can reduce the impact of haploid gene action and act toward preservation of gametic equality, they do not necessarily always do so. Their existence is not an argument against functional differentiation of spermatids by virtue of haploid gene action; however, the cytoplasmic location and diffusibility of the gene product must be known prior to assessing any possible effect of cytoplasmic bridges.

6.3.2 Time of Gene Action

Many of the genes discussed above in Section 6.2, which are actively transcribed during the haploid phase of development, are also transcribed, albeit at a lower level, during meiosis. This may be the case for the testis-specific α-tubulin, and possibly for the protamines, and certainly true for PGK-2. Furthermore, in many of the studies enumerated above, the limits of resolution are simply insufficient to eliminate the possibility of meiotic transcription (complicating factors being both the purity of the cell fractions used and the limits of detection by probe

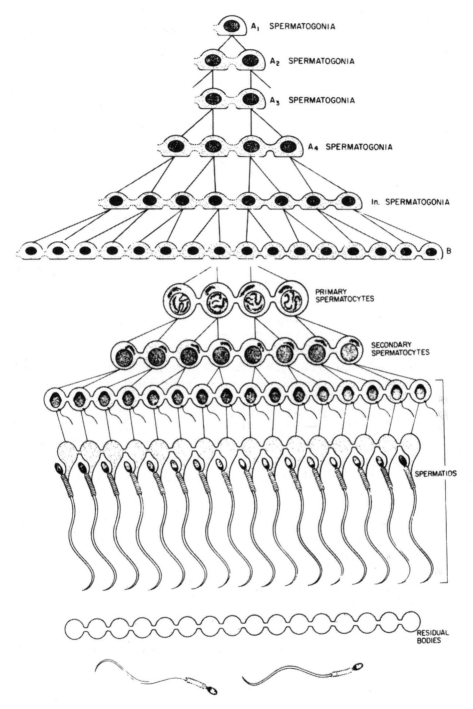

Fig. 5. The relationships of germ cells interconnected by intercellular bridges. Spermatozoa are separated just prior to their release from the seminiferous epithelium. Cells are drawn in linear array for simplicity. (Dym and Fawcett 1971)

hybridization). If transcription is not restricted to the haploid phase, there may be no differences between meiotic segregants, and therefore no biological consequence of haploid transcription.

For gene products newly synthesized during spermiogenesis, templates can either be synthesized (haploid gene action) or activated (haploid gene expression) during spermiogenesis. Analysis of transcription and translation of both PGK-2 (Gold et al. 1983a) and the mouse protamine 1 and 2 messages (Kleene et al. 1984) provide evidence for translational-level regulation of the expression of these two genes. Storage of templates synthesized during meiosis and their subsequent activation for translation may be strategy used for those products needed only during spermiogenesis.

6.3.3 Genetic Evidence

As discussed in detail previously (Sect. 2.3), the property of meiotic segregation exhibited by t-haplotype chromosomes is the firmest evidence for haploid gene action producing two functionally different classes of sperm cells. Much evidence – Braden's (1958) delayed mating effect, the demonstration by Silver and Olds-Clarke (1984) that two populations of sperm are deposited in the female tract by $t/+$ males, and the data of Olds-Clarke and Peitz (1985) showing differences in fertilization among sperm from congenic $+/+$ and $t/+$ males – argues cogently for the production of two functionally different sperm populations from $t/+$ mice. This is presumably by virtue of the haploid action of one or more genes specifying a product influencing some aspect of sperm function.

The other genetic evidence for meiotic segregation distortion in mice is from an unusual line of transgenic mice (Palmiter et al. 1984). These mice were produced by the creation of a transgenic mouse by injection of two copies of pMK (a metallothionein-thymidine kinase fusion gene) oriented as inverted repeats. The original mouse was a mosaic; however, the pMK insert has been stably transmitted only by females. Males carrying the pMK insert are fertile; however, they never transmit the pMK insert to their offspring. Analysis of DNA of sperm obtained from the cauda epididymis suggests that less than half of the sperm carrying the pMK insert fail to reach the cauda; failure to transmit the insert suggests that even those pMK-carrying sperm that do reach the cauda and vas deferens are infertile. The authors proposed that the pMK gene disrupted an essential gene expressed only during the haploid phase of spermatogenesis.

However, these two cases, t-haplotype chromosomes and the pMK insert of transgenic mice, are the only genetic examples of some form of distorted transmission. None of the autosomal recessive genes causing male sterility discussed in Section 2.1.1 exhibit meiotic segregation distortion in the heterozygous state. If these genes specified protein products essential to sperm that were synthesized from templates transcribed during spermiogenesis, we would expect reduced transmission of the mutant alleles from heterozygous males. Not only is this not the case, there is no evidence for any phenotypic effect on male fertility resulting from the heterozygous state for these genes. This does not negate the action of these genes during spermatogenesis, it simply suggests that the action of these genes is not during the haploid phase.

Further evidence that haploid gene action may be rare or insignificant in the determination of the sperm phenotype comes from studies of chromosomally unbalanced gametes. Ford (1972) has examined offspring both from crosses involving translocation heterozygotes and from crosses between hybrids with frequent nondisjunction. The frequency of aneuploid embryos from these crosses was concordant with the frequency expected if chromosomally unbalanced spermatozoa achieved fertilization in proportion to the frequency with which they were produced. These results suggest that gross genome imbalance of sperm is not detrimental to the fertilizing ability of those sperm, implying that haploid action of deleted or duplicated genetic regions does not determine sperm phenotype. Similar conclusions were reached from a study of the fertilizing ability of sperm lacking the H-2 and T-locus (Lyon et al. 1972). This showed that neither of these loci has haploid action essential to fertilization, but it does not exclude the possibility of haploid action not required for fertilization.

Thus, available genetic evidence suggests that transmission distortion, as would be expected from haploid gene action, is rare. This, in turn, implies either that haploid gene action is rare or that mechanisms exist to preserve gametic equality.

7 Resumé and Prospectus

In this chapter I have deliberately juxtaposed information about genes and chromosome aberrations known to affect the process of spermatogenesis with information about the molecular control of spermatogenesis. The existence of some dichotomy between these areas of research is a reflection of the fact that there has been insufficient use of mouse genetic models in the molecular and cellular analysis of spermatogenesis.

Mouse autosomal recessive genes that affect the process of spermiogenesis are all pleiotropic and, therefore, unlikely to specify products unique to spermatogenesis. Since the autosomal recessive mutations cause disruption of spermatogenesis during the haploid phase of differentiation, mice bearing these mutations are of potential value in understanding the importance of haploid gene transcription as well as in unraveling the complex differentiative events of spermiogenesis. Knowledge of the primary gene effect, or, alternatively, gene probes to analyze gene products are likely to be essential in determining the mode of action of these autosome mutations.

I have also discussed chromosome translocations that result in the interruption of spermatogenesis primarily during meiosis. Our understanding of both the events of meiotic prophase and the mechanism of disruption of spermatogenesis by chromosome translocations will be furthered by a thorough biochemical and molecular analysis of DNA synthesis, RNA transcription, and changes in chromatin structure during meiosis. As we gain more complete knowledge of these events, translocation-sterile mice may well be useful in defining regulatory and control points during meiosis.

It is my hope that the information and ideas presented in this chapter will stimulate more use of mouse male-sterile mutants in the experimental analysis of spermatogenesis. It is through this route that we are most likely to gain information leading to understanding the genetic regulation of spermatogenesis.

Acknowledgments. I am deeply appreciative of helpful comments and discussions with many colleagues, in particular, T. Ashley, L. Beckers, P. de Boer, E. Eicher, N. Ganguly, I. Greenblatt, M. Griswold, N. Hecht, W. Hennig, K. Jeon, H. Kremer, P. Olds-Clarke, C. Park, L. B. Russell, G. Rinchik, and R. Sotomayor. I owe an immense debt of gratitude to Margaret Green for her careful and thorough work in compiling Genetic Variants and Strains of the Laboratory Mouse, which was a constant and invaluable reference source for the preparation of this chapter. I thank M. Dawson and P. W. Lane for assistance in obtaining data presented in Table 1, and L. Lameier, C. Lynn, and C. Park for their help in the preparation of the manuscript. Work in my laboratory and the writing of this chapter have been funded in part by the NIH, #HD 16978 and a University of Tennessee Faculty Leave Award.

References

Artzt K, Shin HS, Bennett D (1982) Gene mapping within the T/t-complex of the mouse. II. Anomalous position of the H-2 complex in t-haplotypes. Cell 28:471–476

Ashley T (1983) Nonhomologous synapsis of the sex chromosomes in the heteromorphic bivalents of two X-7 translocations in male mice: R5 and R6. Chromosoma (Berl) 88:178–183

Ashley TA, Cacheiro NLA, Russell LB (1980a) Assessment of factors involved in initiation of synapsis in three X-7 translocations in male mice. Oxford Chromosome Conference, Oxford, England

Ashley T, Cacheiro NLA, Russell LB (1980b) Synaptonemal complex analysis of three X-7 translocations in male mice: Assessment of factors involved in initiation of synapsis. Second International Congress of Cell Biology, Berlin, Germany

Ashley T, Russell LB, Cacheiro NLA (1982) Synaptonemal complex analysis of two X-7 translocations in male mice: R3 and R5. Chromosoma (Berl) 87:149–164

Ashley T, Russell LB, Cacheiro NLA (1983) Synaptonemal complex analysis of two X-7 translocations in male mice: R2 and R6 quadrivalents. Chromosoma (Berl) 88:171–177

Baranov VS, Dyban AP (1968) Analysis of spermatogenic and embryogenic abnormalities in mice heterozygous for the chromosome translocation T6. Genetika 4(12):70–83

Baumann NA, Bourre JM, Jaque C, Pollet S (1972) Genetic disorders of myelination. In: Ciba Found Symp, Lipids, Malnutrition and the Developing Brain. Elsevier, Amsterdam, North-Holland, pp 91–100

Beatty RA (1970) The genetics of the mammalian gamete. Biol Rev 45:73–119

Beatty RA (1972) The genetics of size and shape of spermatozoan organelles. In: Beatty RA, Gluecksohn-Waelsch S (eds) The genetics of the spermatozoon. University of Edinburgh, Edinburgh, pp 97–115

Beechey CV (1973) X-Y Chromosome dissociation and sterility in the mouse. Cytogenet Cell Genet 12:60–67

Beechey CV, Kirk M, Searle AG (1980) A reciprocal translocation induced in an oocyte and affecting fertility in male mice. Cytogenet Cell Genet 27:129–146

Bellve AR (1979) The molecular biology of mammalian spermatogenesis. In: Finn CA (ed) Oxford reviews of reproductive biology. Oxford University Press (Clarendon), London New York, pp 159–261

Bellve AR (1982) Biogenesis of the mammalian spermatozoon. In: Amann RP, Seidel GE (eds) Prospects for sexing mammalian sperm. Colorado Associated University Press, Boulder, pp 69–102

Bellve AR, O'Brien DA (1983) The mammalian spermatozoon: structure and temporal assembly. In: Hartman JF (ed) Mechanism and control of animal fertilization. Academic Press, London, pp 55–137

Bellve AR, Millette CF, Bhatnagar YM, O'Brien DA (1977) Dissociation of the mouse testis and characterization of isolated spermatogenic cells. J Histochem Cytochem 25:480–494

Bennett D (1975) The *T*-locus of the mouse. Cell 6:441–454

Bennett D, Dunn LC (1967) Studies of effects of *t*-alleles in the house mouse on spermatozoa. I. Male sterility effects. J Reprod Fertil 13:421–428

Bennett WI, Gall AM, Southard JL, Sidman RL (1971) Abnormal spermiogenesis in quaking, a myelin-deficient mutant mouse. Biol Reprod 5:30–58

Bhatnagar YM, Romrell LJ, Bellve AR (1985) Biosynthesis of specific histones during meiotic prophase of mouse spermatogenesis. Biol Reprod 32:599–609

Bishop C, Weissenbach J, Casanova M, Bernheim A, Fellous M (1985) DNA sequences and analysis of the human Y chromosome. In: Sandberg AA (ed) The Y chromosome, part A: basic characteristics of the Y chromosome. Alan R Liss, New York, pp 141–176

Boitani C, Geremia R, Rossi R, Monesi V (1980) Electrophoretic pattern of polypeptide synthesis in spermatocytes and spermatids of the mouse. Cell Differ 9:41–49

Bonhomme F, Guenet J-L, Catalan J (1982) Presence d'un facteur de sterilite male, *Hst-2*, segregant dans les croisements interspecifiques. CR Acad Sci Paris 294 Ser III:691–693

Braden AW (1958) Influence of time of mating on the segregation ratio of alleles at the T locus in the house mouse. Nature 181:786–787

Bryan JHD (1977a) Spermatogenesis revisited. III. The course of spermatogenesis in a male-sterile pink-eyed mutant type in the mouse. Cell Tissue Res 180:173–186

Bryan JHD (1977b) Spermatogenesis revisited. IV. Abnormal spermiogenesis in mice homozygous for another male-sterility-inducing mutation *hpy* (hydrocephalic-polydactyl). Cell Tissue Res 180:187–201

Bryan JHD, Chandler DB (1978) Tracheal ciliary defects in mice homozygous for a recessive pleiotropic mutation hydrocephalic-polydactyl. J Cell Biol 79:281a

Burgoyne PS (1975) Sperm phenotype and its relationship to somatic and germ line genotype: a study using mouse aggregation chimeras. Dev Biol 44:63–76

Burgoyne PS (1978) The role of the sex chromosomes in mammalian germ cell differentiation. Ann Biol Anim Biochim Biophys 18(2B):317–325

Burgoyne PS (1979) Evidence for an association between univalent Y chromosomes and spermatocyte loss in XYY mice and men. Cytogenet Cell Genet 23:84–89

Burgoyne PS, Baker TG (1984) Meiotic pairing and gametogenic failure. In: Evans CW, Dickinson HG (eds) 38th Symp Soc Exp Biol Controlling events in meiosis. Company of Biologists, Cambridge, pp 349–362

Burgoyne PS, Biddle FG (1980) Spermatocyte loss in XYY mice. Cytogenet Cell Genet 28:143–144

Burgoyne PS, Levy ER, McLaren A (1986) Spermatogenic failure in male mice lacking H-Y antigen. Nature 320:170–172

Cacheiro NLA (1978) Mouse News Lett 59:46

Cacheiro NLA (1983) Involvement of the Y chromosome in translocations, ORNL Biology Division Annual Report. ORNL 6021:121–122

Cacheiro NLA, Russell LB, Swartout MS (1974) Translocations, the predominant cause of total sterility in sons of mice treated with mutagens. Genetics 76:73–91

Cacheiro NLA, Russell LB, Bangham JW (1978) A new mouse X-autosome translocation with nonrandom inactivation. Genetics 88:s13–s14

Cattanach BM (1961a) A chemically-induced variegated-type position effect in the mouse. Z Vererbungsl 92:165–182

Cattanach BM (1961b) XXY mice. Genet Res 2:156–160

Cattanach BM (1966) The location of Cattanach's translocation in the X-chromosome linkage map of the mouse. Genet Res 8:253–256

Cattanach BM (1974) Position effect variegation in the mouse. Genet Res 23:291–306

Cattanach BM (1975) Sex reversal in the mouse and other mammals. In: Ball M, Wild AE (eds) The early development of mammals. Cambridge University Press, Cambridge, pp 305–317

Cattanach BM, Kirk M (1985) Differential activity of maternally and paternally derived chromosome regions in mice. Nature 315:496–498
Cattanach BM, Pollard CE (1969) An XYY sex-chromosome constitution in the mouse. Cytogenet 8:80–86
Cattanach BM, Pollard CE, Hawkes SG (1971) Sex-reversed mice: XX and XO males. Cytogenet 10:318–337
Chandley AC (1982) A pachytene analysis of two male-fertile paracentric inversions in Chromosome 1 of the mouse and in the male-sterile double heterozygote. Chromosoma (Berl) 85:127–135
Chandley AC (1984) Infertility and chromosome abnormality. Oxford Rev Reprod Biol 6:1–46
Chubb C, Nolan C (1985) Animal models of male infertility: mice bearing single-gene mutations that induce infertility. Endocrinology 117:338–346
Clarke GN, Yanagimachi R (1978) Actin in mammalian sperm heads. J Exp Zool 205:125–132
Coleman DL (1962) Effect of genic substitution on the incorporation of tyrosine into the melanin of mouse skin. Arch Biochem Biophys 96:562–568
Coniglio JB, Grogan WM, Harris DG, Fitzhugh ML (1975) Lipid and fatty acid composition of testes of quaking mice. Lipids 10:109–112
Coyne JA (1985) The genetic basis of Haldane's rule. Nature 314:736–738
D'Agostino A, Geremia R, Monesi V (1978) Post-meiotic gene activity in spermatogenesis of the mouse. Cell Differ 7:175–183
Danska J, Silver LM (1980) Cell-free translation of a T/t complex cell surface associated gene product. Cell 22:901–904
Das RK, Kar RN (1981) A 41,XYY mouse. Experientia (Basel) 37:821–822
de Boer P (1976) Male meiotic behavior and male and female litter size in mice with the T(2;8)26H and T(1;13)7OH reciprocal translocations. Genet Res 27:369–387
de Boer P (1986) Chromosomal causes for fertility reduction in mammals. In: de Serres FJ (ed) Chemical mutagens, vol 10. Plenum, New York, pp 427–467
de Boer P, Branje HEB (1979) Association of the extra chromosome of tertiary trisomic male mice with the sex chromosome during first meiotic prophase and its significance for impairment of spermatogenesis. Chromosoma (Berl) 73:369–379
de Boer P, Groen A (1974) Fertility and meiotic behavior of male T70H tertiary trisomics of the mouse. Cytogenet Cell Genet 13:489–510
de Boer P, Nijhoff JH (1981) Incomplete sex chromosome pairing in oligospermic male hybrids of *Mus musculus* and *M. musculus molossinus* in relation to the source of the Y chromosome and the presence or absence of a reciprocal translocation. J Reprod Fertil 62:235–243
de Boer P, van der Hoeven FA (1977) Son-sire regression based on heritability estimates of chiasma frequency, using T70H mouse translocation heterozygotes, and the relation between univalence, chiasma frequency and sperm production. Heredity 39:335–343
de Boer P, Searle AG, van der Hoeven FA, de Rooij DG, Beechey CV (1986) Male pachytene pairing in single and double translocation heterozygotes and spermatogenic impairment in the mouse. Chromosoma (Berl) 93:326–336
Distel RJ, Kleene KC, Hecht NB (1984) Haploid expression of a mouse testis alpha-tubulin gene. Science 224:68–70
Dooher GB, Bennett D (1977) Spermiogenesis and spermatozoa in sterile mice carrying different lethal T/t locus haplotypes: a transmission and scanning electron microscopic study. Biol Reprod 17:269–288
Duchen LW, Strich SJ, Falconer DS (1968) An hereditary motor neuron disease with progressive denervation of muscle in the mouse: the mutant "Wobbler". J Neurol Neurosurg Psychiatry 31:535–542
Dudley K, Potter J, Lyon MF, Willison KR (1984) Analysis of male sterile mutations in the mouse using haploid stage expressed cDNA probes. Nucl Acids Res 12:4281–4293
Dym M, Fawcett DW (1971) Further observations on the numbers of spermatogonia, spermatocytes and spermatids connected by intercellular bridges in the mammalian testis. Biol Reprod 4:195–215
Eicher EM (1970) X-Autosome translocations in the mouse: total inactivation versus partial inactivation of the X-chromosome. Adv Genet 15:175–259

Eicher EM (1982) Primary sex determining genes in mice. In: Amann RP, Seidel GE (eds) Prospects for sexing mammalian sperm. Colorado Associated University Press, Boulder, pp 121–135

Eicher EM, Washburn LL (1977) Mouse News Lett 56:43

Eicher EM, Washburn LL (1978) Assignment of genes to regions of mouse chromosomes. Proc Natl Acad Sci USA 75:946–950

Eicher EM, Washburn LL (1986) Genetic control of primary sex determination in mice. Ann Rev Genet 20:327–360

Eicher EM, Cherry M, Flaherty L (1978) Autosomal phosphoglycerate kinase linked to mouse major histocompatibility complex. Mol Gen Genet 158:225–228

Eicher EM, Phillips SJ, Washburn LL (1983) The use of molecular probes and chromosomal rearrangements to partition the mouse Y chromosome into functional regions. In: Messer A, Porter IH (eds) Recombinant DNA and medical genetics. Academic Press, London, pp 57–71

Erickson RP (1973) Haploid gene expression versus meiotic drive: the relevance of intercellular bridges during spermatogenesis. Nature New Biol 243:210–212

Erickson RP (1984) Cattanach's translocation [Is(7:X)Ct] corrects male sterility due to homozygosity for Chromosome 7 deletions. Genet Res 43:35–41

Erickson RP, Erickson JM, Betlach CJ, Meistrich ML (1980) Further evidence for haploid gene expression during spermatogenesis: heterogeneous, poly(A)-containing RNA is synthesized post-meiotically. J Exp Zool 214:13–19

Evans EP (1976) Male sterility and double heterozygosity for Robertsonian translocations in mouse. Chromosomes Today 5:75–81

Evans EP, Ford CE, Searle AG (1969) A 39,X/41,XYY mosaic mouse. Cytogenet 8:87–96

Evans EP, Beechey CV, Burtenshaw MD (1978) Meiosis and fertility in XYY mice. Cytogenet Cell Genet 20:249–263

Evans EP, Burtenshaw MD, Cattanach BM (1982) Meiotic crossing over between the X and Y chromosomes of male mice carrying the sex-reversing (*Sxr*) factor. Nature 300:443–445

Fawcett DW (1975) The mammalian spermatozoon. Dev Biol 44:394–436

Fawcett DW, Anderson WA, Phillips DM (1971) Morphogenetic factors influencing the shape of the sperm head. Dev Biol 26:220–251

Feig LA, Bellve AR, Erickson NH, Klagsbrun M (1980) Sertoli cells contain a mitogenic polypeptide. Proc Natl Acad Sci USA 77:4774–4778

Ford CE (1972) Gross genome unbalance in mouse spermatozoa: does it influence the capacity to fertilize? In: Beatty RA, Gluecksohn-Waelsch S (eds) Proc Int Symp Genetics of the Spermatozoan. University of Edinburgh, Edinburgh, pp 359–369

Forejt J (1976) Spermatogenic failure of translocation heterozygotes affected by H-2 linked gene in mouse. Nature 260:143–145

Forejt J (1978) Cited as personal communication. In: Green MC (ed) (1981) Genetic variants and strains of the laboratory mouse. Gustav Fischer, Stuttgart

Forejt J (1982) X-Y involvement in male sterility caused by autosome translocations – a hypothesis. In: Crosignani PG, Rubin BL (eds) Genetic control of gamete production and function. Academic Press/Grune and Stratton, New York, pp 135–151

Forejt J, Gregorova S (1977) Meiotic studies of translocations causing male sterility in the mouse. I. Autosomal reciprocal translocations. Cytogenet Cell Genet 19:159–179

Forejt J, Ivanyi P (1975) Genetic studies on male sterility of hybrids between laboratory and wild mice (*Mus musculus* L.). Genet Res 24:189–206

Forejt J, Gregorova S, Capkova A (1979) Restoration of fertility in males carrying T43H malesterile translocation. Mouse News Lett 61:50

Forejt J, Capkova J, Gregorova S (1980) T(16;17)43H translocation as a tool in analysis of the proximal part of Chromosome 17 (including *T-t* gene complex) of the mouse. Genet Res 35:165–177

Forsthoefel PF (1962) Genetics and manifold effects of Strong's luxoid gene in the mouse, including its interactions with Green's luxoid and Carter's luxate genes. J Morphol 110:391–420

Fujimoto H, Erickson RP (1982) Functional assays for mRNA detect many new messages after male meiosis in mice. Biochem Biophys Res Commun 108:1369–1375

Fujimoto H, Erickson RP, Quinto M, Rosenberg MP (1984) Postmeiotic transcription in mouse testes detected with spermatid cDNA clones. Biosci Rep 4:1037–1044

Gizang-Ginsberg E, Wolgemuth DJ (1985) Localization of mRNA's in mouse testes by in situ hybridization: distribution of α-tubulin and developmental stage specificity of pro-opiomelanocortin transcripts. Dev Biol 111:293–305

Gold B, Fujimoto H, Kramer JM, Erickson RP, Hecht NB (1983a) Haploid accumulation and translational control of phosphoglycerate kinase-2 messenger RNA during mouse spermatogenesis. Dev Biol 98:392–399

Gold B, Stern L, Bradley FM, Hecht NB (1983b) Gene expression during mammalian spermatogenesis. II. Stage-specific differences in mRNA populations. J Exp Zool 225:123–134

Goldberg E, Sberna D, Wheat TE, Urbanski GJ, Margoliash E (1977) Cytochrome C: immunofluorescent localization of the testis-specific form. Science 196:1010–1012

Green MC (1964) Mouse News Lett 30:32

Green MC (ed) (1981) Genetic variants and strains of the laboratory mouse. Gustav Fischer, Stuttgart

Gregorova S, Baranov VS, Forejt J (1981) Partial trisomy (including T-t gene complex) of the Chromosome 17 of the mouse. The effect on male fertility and the transmission to progeny. Folia Biol (Prague) 27:171–177

Griswold MD (1987) Protein secretions of Sertoli cells. Int Rev Cytol (in press)

Gropp A, Winking H (1981) Robertsonian translocations: cytology, meiosis, segregation patterns and biological consequences of heterozygosity. Symp Zool Soc Lond 47:141–181

Gropp A, Winking H, Redi C (1982) Consequences of Robertsonian heterozygosity: segregational impairment of fertility versus male-limited sterility. In: Crosignani PG, Rubin BL (eds) Genetic control of gamete production and function. Grune & Stratton, New York, pp 115–134

Gummere GR, McCormick PJ, Bennett D (1986) The influence of genetic background and the homologous Chromosome 17 on t-haplotype transmission ratio distortion in mice. Genetics 114:235–245

Haldane JBS (1922) Sex ratio and unisexual sterility in hybrid animals. J Genet 12:101–109

Handel MA (1985) Mouse News Lett 72:124

Handel MA, Dawson M (1981) Effects on spermiogenesis in the mouse of a male sterile neurological mutation, Purkinje cell degeneration. Gamete Res 4:185–192

Handel MA, Washburn LL, Rosenberg MP, Eicher EM (1987) Male sterility caused by p^{6H} and qk mutations is not corrected in chimeric mice. J Exp Zool 243:81–91

Hash DC, Wolfe HG (1979) Pink-eyed dilution alleles affect negative surface charges of mouse spermatozoa. Dev Genet 1:61–68

Hecht NB, Kleene KC, Distel RJ, Silver LM (1984) The differential expression of the actins and tubulins during spermatogenesis in the mouse. Exp Cell Res 153:275–279

Hecht NB, Bower PA, Kleene KC, Distel RJ (1985) Size changes of protamine 1 mRNA provide a molecular marker to monitor spermatogenesis in wild type and mutant mice. Differentiation 29:189–193

Hecht NB, Bower PA, Waters SH, Yelick PC, Distel RJ (1986) Evidence for haploid expression of mouse testicular genes. Exp Cell Res 164:183–190

Hennig W (1985) Y chromosome function and spermatogenesis in Drosophila hydei. Adv Genet 23:179–234

Herrmann B, Bucan M, Mains PE, Frischauf A-M, Silver LM, Lehrach H (1986) Genetic analysis of the proximal portion of the mouse t complex: evidence for a second inversion within t haplotypes. Cell 44:469–476

Hillman N, Nadijcka M (1978a) A comparative study of spermiogenesis in wild-type and T:t-bearing mice. J Embryol Exp Morphol 44:243–261

Hillman N, Nadijcka M (1978b) A study of spermatozoan defects in wild-type and T:t-bearing mice. J Embryol Exp Morphol 44:263–280

Hillman N, Nadijcka M (1980) Sterility in mutant (t^{Lx}/t^{Ly}) male mice. I. A morphological study of spermiogenesis. J Embryol Exp Morphol 59:27–37

Hogg H, McLaren A (1985) Absence of a sex vesicle in meiotic foetal germ cells is consistent with an XY sex chromosome constitution. J Embryol Exp Morphol 88:327–332

Hollander WF (1976) Hydrocephalic-polydactyl, a recessive and pleiotropic mutant in the mouse, and its location in Chromosome 6. Iowa State J Res 51:13–23

Hollander WF, Bryan JHD, Gowen JW (1960) A male sterile pink-eyed mutant type in the mouse. Fertil Steril 11:316–324

Hotta Y, Chandley AC (1982) Activities of X-linked enzymes in spermatocytes of mice rendered sterile by chromosomal alterations. Gamete Res 6:65–72

Hotta Y, Ito M, Stern H (1966) Synthesis of DNA during meiosis. Proc Natl Acad Sci USA 56:1184–1191

Hotta Y, Chandley AC, Stern H, Searle AG, Beechey CV (1979) A disruption of pachytene DNA metabolism in male mice with chromosomally derived sterility. Chromosoma (Berl) 73:287–300

Hotta Y, Tabata S, Stubbs L, Stern H (1985) Meiosis-specific transcripts of a DNA component replicated during chromosome pairing: homology across the phylogenetic spectrum. Cell 40:785–793

Huckins C, Bullock LP, Long JL (1981) Morphological profiles of cryptorchid XXY mouse testes. Anat Rec 199:507–518

Hugenholtz AP (1984) Mouse News Lett 71:34–35

Hunt DM, Johnson DR (1971) Abnormal spermiogenesis in two pink-eyed sterile mutants in the mouse. J Embryol Exp Morphol 26:111–121

Imai HT, Matsuda Y, Shiroishi T, Moriwaki K (1981) High frequency of X-Y chromosome dissociation in primary spermatocytes of F_1 hybrids between Japanese wild mice (*Mus musculus molossinus*) and inbred laboratory mice. Cytogenet Cell Genet 29:166–175

Johnson AD (1970) Testicular lipids. In: Johnson AD, Gomes WR, Vandemark NL (eds) The testis, vol II. Biochemistry. Academic Press, London, pp 193–258

Johnson DR, Hunt DM (1971) Hop-sterile, a mutant gene affecting sperm tail development in the mouse. J Embryol Exp Morphol 25:223–236

Johnson DR, Hunt DM (1975) Endocrinological findings in sterile pink-eyed mice. J Reprod Fertil 42:51–58

Keitges E, Rivest M, Siniscalco M, Gartler SM (1985) X-linkage of steroid sulphatase in the mouse is evidence for a functional Y-linked allele. Nature 315:226–227

Kierszenbaum AL, Tres LL (1974a) Nucleolar and perichromosomal RNA synthesis during meiotic prophase in the mouse testis. J Cell Biol 60:39–53

Kierszenbaum AL, Tres LL (1974b) Transcription sites in spread meiotic prophase chromosomes from mouse spermatocytes. J Cell Biol 63:923–935

Kleene KC, Distel RJ, Hecht NB (1983) cDNA clones encoding cytoplasmic poly(A)+ RNA, which first appear at detectable levels in haploid phases of spermatogenesis in the mouse. Dev Biol 98:455–464

Kleene KC, Distel RJ, Hecht NB (1984) Translational regulation and deadenylation of a protamine mRNA during spermiogenesis in the mouse. Dev Biol 105:71–79

Kleene KC, Distel RJ, Hecht NB (1985) Nucleotide sequence of a cDNA clone encoding mouse protamine 1. Biochem 24:719–722

Kofman-Alfaro S, Chandley AC (1970) Meiosis in the male mouse. An autoradiographic investigation. Chromosoma (Berl) 31:404–426

Kot MC, Handel MA (1987) Binding of morphologically abnormal sperm to mouse egg zonae pellucidea. Gam Res (in press)

Kramer JM (1981) Immunofluorescent localization of PGK-1 and PGK-2 isozymes within specific cells of the mouse testis. Dev Biol 87:30–36

Kramer JM, Erickson RP (1981) Developmental program of PGK-1 and PGK-2 isozymes in spermatogenic cells of the mouse: specific activities and rates of synthesis. Dev Biol 87:37–45

Kramer JM, Erickson RP (1982) Analysis of stage-specific protein synthesis during spermatogenesis of the mouse by two-dimensional gel electrophoresis. J Reprod Fertil 64:139–144

Krzanowska H (1972) Influence of Y chromosome on fertility in mice. In: Beatty RA, Gluecksohn-Waelsch S (eds) Proc Int Symp Genetics of the Spermatozoon. University of Edinburgh, Edinburgh, pp 370–386

Krzanowska H (1974) The passage of abnormal spermatozoa through the uterotubal junction of the mouse. J Reprod Fertil 38:81–90

Krzanowska H (1986) Interstrain competition amongst mouse spermatozoa inseminated in various proportions, as affected by the genotype of the Y chromosome. J Reprod Fertil 77:265–270

Krzanowska H, Lorenc E (1983) Influence of egg investments on *in vitro* penetration of mouse eggs by misshapen spermatozoa. J Reprod Fertil 68:57–62

LaCroix M, Parvinen M, Fritz JB (1981) Localization of testicular plasminogen activator in discrete portions (stages VII and VIII) of the seminiferous tubule. Biol Reprod 25:143–146

Landis SC, Mullen RJ (1978) The development and degeneration of Purkinje cells in *pcd* mutant mice. J Comp Neurol 177:125–143

Leblond CP, Clermont Y (1952) Spermiogenesis of rat, mouse, hamster and guinea pig as revealed by the "periodic acid-fuchsin sulfurous acid" technique. Am J Anat 90:167–216

Leestma JE, Sepsenwol S (1980) Sperm tail axoneme alterations in the Wobbler mouse. J Reprod Fertil 58:267–270

Leonard A, Deknudt G (1969) Etude cytologique d'une translocation chromosome Y-autosome chez souris. Experientia (Basel) 25:876–877

Lewis SE, Turchin HA, Wojtowicz TE (1978) Fertility studies of complementing genotypes at the albino locus of the mouse. J Reprod Fertil 53:197–202

Lifschytz E, Lindsley DL (1972) The role of X-chromosome inactivation during spermatogenesis. Proc Natl Acad Sci USA 69:182–186

Liming S, Pathak S (1981) Gametogenesis in a male Indian muntjac X Chinese muntjac hybrid. Cytogenet Cell Genet 30:152–156

Lopez LC, Bayna EM, Litoff D, Shaper NL, Shur BD (1985) Receptor function of mouse sperm surface galactosyltransferase during fertilization. J Cell Biol 101:1501–1510

Lyon MF (1981) The *t*-complex and the genetical control of development. Symp Zool Soc Lond 47:455–477

Lyon MF (1984) Transmission ratio distortion in mouse *t*-haplotypes is due to multiple distorter genes acting on a responder locus. Cell 37:621–628

Lyon MF (1986) Male sterility of the mouse *t*-complex is due to homozygosity of the distorter genes. Cell 44:357–363

Lyon MF, Meredith R (1966) Autosomal translocations causing male sterility and viable aneuploidy in the mouse. Cytogenet 5:335–354

Lyon MF, Searle AG, Ford CE, Ohno S (1964) A mouse translocation suppressing sex-linked variegation. Cytogenet 3:306–323

Lyon MF, Glenister PH, Hawker SG (1972) Do the *H-2* and *T*-loci of the mouse have a function in the haploid phase of sperm? Nature 240:152–153

Lyon MF, Glenister PH, Lamoreux ML (1975) Normal spermatozoa from androgen-resistant germ cells of chimaeric mice and the role of androgen in spermatogenesis. Nature 258:620–622

Mahadevaiah S, Mittwoch V, Moses MJ (1984) Pachytene chromosomes in male and female mice heterozygous for the Is(7;1)40H insertion. Chromosoma (Berl) 90:163–169

Matsuda Y, Imai HT, Moriwaki K, Kondo K, Bonhomme F (1982) X-Y chromosome dissociation in wild derived *Mus musculus* subspecies, laboratory mice, and their F_1 hybrids. Cytogenet Cell Genet 34:241–252

Matsuda Y, Imai HT, Moriwaki K, Kondo K (1983) Modes of inheritance of X-Y dissociation in inter-subspecies hybrids between BALB/c mice and *Mus musculus molossinus*. Cytogenet Cell Genet 35:209–215

McGrath J, Hillman N (1980) Sterility in mutant (t^{Lx}/t^{Ly}) male mice. III. *In vitro* fertilization. J Embryol Exp Morphol 59:49–58

McGrath J, Solter D (1984) Completion of mouse embryogenesis requires both the maternal and paternal genomes. Cell 37:179–183

McLaren A (1976) Mammalian chimaeras. Cambridge University Press, Cambridge

McLaren A, Simpson E, Tomonari K, Chandler P, Hogg H (1984) Male sexual differentiation in mice lacking H-Y antigen. Nature 312:552–555

Melvold RW (1974) The effects of mutant *p*-alleles on the reproductive system in mice. Genet Res 23:319–325

Miklos GLG (1974) Sex-chromosome pairing and male fertility. Cytogenet Cell Genet 13:558–577

Mittwoch U, Mahadevaiah S, Olive MS (1981) Retardation of ovarian growth in male-sterile mice carrying an autosomal translocation. J Med Genet 18:414–417

Monesi V, Geremia R, D'Agostino A, Boitani C (1978) Biochemistry of male germ cell differentiation in mammals: RNA synthesis in meiotic and post meiotic cells. Curr Top Dev Biol 12:11–36

Moutier R (1976) New mutations causing sterility restricted to the male in rats and mice. In: Antikatzides T, Erichsen S, Spiegel A (eds) The laboratory animal in the study of reproduction. Gustav Fischer, Stuttgart, pp 115–117

Mullen RJ (1977) Site of *pcd* gene action and Purkinje cell mosaicism in cerebella of chimaeric mice. Nature 270:245–247

Mullen RJ (1978) Mosaicism in the central nervous system of mouse chimeras. In: Subtelny S, Sussex IM (eds) The clonal basis of development. Academic Press, London, pp 83–101

Mullen RJ, Eicher EM, Sidman RL (1976) Purkinje cell degeneration, a new neurological mutation in the mouse. Proc Natl Acad Sci USA 73:208–212

Nadijcka M, Hillman N (1980) Sterility in mutant (t^{Lx}/t^{Ly}) male mice. II. A morphological study of spermatozoa. J Embryol Exp Morphol 59:39–47

Nestor A, Handel MA (1984) Transport of morphologically abnormal sperm in the female reproductive tract of mice. Gamete Res 10:119–125

Oakberg EF (1956) A description of spermiogenesis in the mouse and its use in analysis of the cycle of the seminiferous epithelium and germ cell renewal. Am J Anat 99:391–413

Olds PJ (1970) Effect of the *T* locus on sperm distribution in the house mouse. Biol Reprod 2:91–97

Olds-Clarke P (1983a) Nonprogressive sperm motility is characteristic of most complete *t* haplotypes in the mouse. Genet Res 42:151–157

Olds-Clarke P (1983b) The nonprogressive motility of sperm populations from mice with a t^{w32} haplotype. J Androl 4:136–143

Olds-Clarke P (1986) Motility characteristics of sperm from the uterus and oviducts of female mice after mating to congenic male differing in sperm transport and fertility. Biol Reprod 34:453–467

Olds-Clarke P, Becker A (1978) The effect of the *T/t* locus on sperm penetration *in vivo* in the house mouse. Biol Reprod 18:132–140

Olds-Clarke P, Carey JE (1978) Rate of egg penetration *in vitro* accelerated by *T/t* locus in the mouse. J Exp Zool 206:323–332

Olds-Clarke P, McCabe S (1982) Genetic background affects expression of *t* haplotype in mouse sperm. Genet Res 40:249–254

Olds-Clarke P, Peitz B (1985) Fertility of sperm from *t/+* mice: evidence that *t*-bearing sperm are dysfunctional. Genet Res 47:49–52

Palmiter RD, Wilkie TM, Chen HY, Brinster RL (1984) Transmission distortion and mosaicism in an unusual transgenic mouse pedigree. Cell 36:869–877

Pathak S, Elder FFB, Maxwell BL (1980) Asynaptic behavior of X and Y chromosomes in the Virginia opossum and the southern pygmy mouse. Cytogenet Cell Genet 26:142–149

Ponzetto C, Wolgemuth DJ (1985) Haploid expression of a unique c-*abl* transcript in the mouse male germ line. Mol Cell Biol 5:1791–1794

Rathenberg R, Muller D (1973) X and Y chromosome pairing and disjunction in a male mouse with an XYY sex-chromosome constitution. Cytogenet Cell Genet 12:87–92

Redi CA, Garagna S, Hilscher B, Winking H (1985) The effects of some Robertsonian chromosome combinations on the seminiferous epithelium of the mouse. J Embryol Exp Morphol 85:1–19

Ritzen EM, Boitani C, Parvinen M, French F, Feldman M (1982) Stage dependent secretion of ABP by rat seminiferous tubules. Mol Cell Endocrinol 25:25–33

Roderick TH (1976) Inversions of mice in studies of mutagenesis. Genetics 83:s64

Roehme D, Fox H, Herrmann B, Frischauf A-M, Edstrom JE, Mains P, Silver LM, Lehrach H (1984) Molecular clones of the mouse *t* complex derived from microdissected metaphase chromosomes. Cell 36:783–788

Russell LB (1963) Mammalian X-chromosome action: inactivation limited in spread and in region of origin. Science 140:976–978

Russell LB (1972) A second T(X;8) in the mouse. Genetics 71:s53–54

Russell LB (1983) X-autosome translocations in the mouse: their characterization and use as tools to investigate gene inactivation and gene action. In: Sandberg AA (ed) Cytogenetics of the mammalian X chromosome, part A. Basic mechanisms of X chromosome behavior. Alan R Liss, New York, pp 205–250

Russell LB, Bangham JW (1959) Variegated-type position effects in the mouse. Genetics 44:532

Russell LB, Bangham JW (1961) Variegated-type position effects in the mouse. Genetics 46:509–525

Russell LB, Chu EHY (1961) An XXY male in the mouse. Proc Nat Acad Sci USA 47:571–575

Russell LB, Montgomery CS (1969) Comparative studies on X-autosome translocations in the mouse. I. Origin, viability, fertility and weight of five T(X;1)'s. Genetics 63:103–120

Russell LB, Montgomery CS (1970) Comparative studies on X-autosome translocations in the mouse. II. Inactivation of autosomal loci, segregation, and mapping of autosomal breakpoints in five T(X;1)'s. Genetics 64:281–312

Russell LB, Cacheiro NLA, Bangham J, Swartout MS (1974) Attempts to locate the X-chromosome inactivation center through the study of new translocations. Annual Report, Biology Division, Oak Ridge National Laboratory. ORNL 4993:121–122

Russell LB, Larsen MM, Cacheiro NLA (1979) Use of chimeras in studying sterility of X-autosome translocation male mice. Genetics 91:s108–109

Russell LB, Cacheiro NLA, Montgomery CS (1980a) Attempts at genetic "rescue" of male sterility resulting from X-autosome translocations. Genetics 94:s89–90

Russell LB, Larsen MM, Cacheiro NLA (1980b) The use of chimeras to determine whether translocations act within germ cells to cause male sterility. Cytogenet Cell Genet 27(4):211

Russell LD (1980) Sertoli-germ cell interrelations: a review. Gamete Res 3:179–202

Sachs L (1954) Sex-linkage and the sex chromosomes in man. Ann Eugen 18:255–261

Samorajski T, Friede RL, Reimer PR (1970) Hypomyelination in the quaking mouse. A model for the analysis of disturbed myelin formation. J Neuropathol Exp Neurol 29:507–523

Searle AG (1974) Nature and consequences of induced chromosome damage in mammals. Genetics 78:173–186

Searle AG (1982) The genetics of sterility in the mouse. In: Crosignani P, Rubin B (eds) Genetic control of gamete production and function. Academic Press, London, pp 93–114

Searle AG, Beechey CV, Evans EP (1978) Meiotic effects in chromosomally derived male sterility of mice. Ann Biol Anim Biochim Biophys 18(2B):391–398

Searle AG, Beechey CV, de Boer P, de Rooij DG, Evans EP, Kirk M (1983a) A male-sterile insertion in the mouse. Cytogenet Cell Genet 36:617–626

Searle AG, Beechey CV, Evans EP, Kirk M (1983b) Two new X-autosome translocations in the mouse. Cytogenet Cell Genet 35:279–292

Sherman MI, Wudl LR (1977) T-complex mutations and their effects. In: Sherman MI (ed) Concepts in mammalian embryogenesis. MIT, Cambridge, USA, pp 136–234

Seitz AW, Bennett D (1985) Transmission distortion of t-haplotypes is due to interactions between meiotic partners. Nature 313:143–144

Shin H-S, Bennett D, Artzt K (1984) Gene mapping within the T/t complex of the mouse. IV: The inverted MHC is intermingled with several t-lethal genes. Cell 39:573–578

Shur BD (1981) Galactosyltransferase activities on mouse sperm bearing multiple t^{lethal} and t^{viable} haplotypes of the T/t complex. Genet Res 38:225–236

Shur BD, Bennett D (1979) A specific defect in galactosyltransferase regulation on sperm bearing mutant alleles of the T/t locus. Dev Biol 71:243–259

Shur BD, Hall NG (1982) A role for mouse sperm surface galactosyltransferase in sperm binding to the egg zona pellucida. J Cell Biol 95:574–579

Sidman RL, Dickie MM, Appel SH (1964) Mutant mice (*quaking* and *jimpy*) with deficient myelination in the central nervous system. Science 144:309–311

Silver LM (1985) Mouse t haplotypes. Ann Rev Genet 19:179–208

Silver LM, Olds-Clarke P (1984) Transmission ratio distortion of mouse t haplotypes is not a consequence of wild-type sperm degeneration. Dev Biol 105:250–252

Silver LM, White M (1982) A gene product of the mouse t complex with chemical properties of a cell surface-associated component of the extracellular matrix. Dev Biol 91:423–430

Silver LM, Artzt K, Bennett D (1979) A major testicular cell protein specified by a mouse T/t complex gene. Cell 17:275–284

Silver LM, Uman J, Danska J, Garrels JJ (1983) A diversified set of testicular cell proteins specified by genes within the mouse *t* complex. Cell 35:35–45

Silvers WK (1979) The coat colors of mice. Springer, Berlin Heidelberg New York

Singh L, Jones KW (1982) Sex reversal in the mouse (*Mus musculus*) is caused by a recurrent nonreciprocal crossover involving the X and an aberrant Y chromosome. Cell 28:205–216

Slizynski BM (1964) Cytology of the XXY mouse. Genet Res 5:328–329

Solari AJ (1970) The spatial relationship of the X and Y chromosomes during meiotic prophase in mouse spermatocytes. Chromosoma (Berl) 29:217–236

Solari AJ (1971) The behaviour of chromsomal axes in Searle's X-autosome translocation. Chromosoma (Berl) 34:99–112

Solari AJ (1974) The behavior of the XY pair in mammals. Int Rev Cytol 38:273–317

Solari AJ, Ashley T (1977) Ultrastructure and behavior of the achiasmatic, telosynaptic XY pair of the sand rat *Psammomys obesus*. Chromosoma (Berl) 62:319–336

Solari AJ, Tres L (1967) The localization of nucleic acids and the argentaffin substance in the sex vesicle of mouse spermatocytes. Exp Cell Res 47:86–96

Sotomayor RE, Handel MA (1986) Failure of acrosome assembly in a male sterile mouse mutant. Biol Reprod 34:171–182

Sotomayor RE, Clark J, Handel MA (1986) Developmental expression of the blind-sterile (*bs*) mutation in mouse testes. Biol Reprod 34(Suppl 1):60

Speed RM (1986) Abnormal RNA synthesis in sex vesicles of tertiary trisomic male mice. Chromosoma (Berl) 93:267–270

Stern L, Gold B, Hecht NB (1983a) Gene expression during mammalian spermatogenesis. I. Evidence for stage-specific synthesis of polypeptides *in vivo*. Biol Reprod 28:483–496

Stern L, Kleene KC, Gold B, Hecht NB (1983b) Gene expression during mammalian spermatogenesis. III. Changes in populations of mRNA during spermiogenesis. Exp Cell Res 143:247–255

Stubbs L, Stern H (1986) DNA synthesis at selective sites during pachytene in mouse spermatocytes. Chromosoma (Berl) 93:529–536

Surani MAH, Barton SC, Norris ML (1984) Development of reconstituted mouse eggs suggests imprinting of the genome during gametogenesis. Nature 308:548–550

Surani MAH, Barton SC, Norris ML (1986) Nuclear transplantation in the mouse: heritable differences between parental genomes after activation of the embryonic genome. Cell 45:127–136

Tanaka S, Fujimoto H (1986) A postmeiotically expressed clone encodes lactate dehydrogenase isozyme X. Biochem Biophys Res Commun 136:760–766

Tessler S, Olds-Clarke P (1981) Male genotype influences sperm transport in female mice. Biol Reprod 24:806–813

Tessler S, Carey JE, Olds-Clarke P (1981) Mouse sperm motility affected by factors in the *T/t* complex. J Exp Zool 217:277–285

Van de Berg JL, Cooper DW, Close PJ (1976) Testis-specific phosphoglycerate kinase. J Exp Zool 198:231–240

Varnum DS (1983) Blind-sterile: a new mutation on Chromosome 2 of the house mouse. J Hered 74:206–207

Venolia L, Cooper DW, O'Brien DA, Millette CF, Gartler SM (1984) Transformation of the HPRT gene with DNA from spermatogenic cells. Chromosoma (Berl) 90:185–189

Waters SH, Distel RJ, Hecht NB (1985) Mouse testes contain two size classes of actin mRNA that are differentially expressed during spermatogenesis. Mol Cell Biol 5:1649–1654

Wieben ED (1981) Regulation of the synthesis of lactate dehydrogenase-X during spermatogenesis in the mouse. J Cell Biol 88:492–498

Willison KR, Dudley K, Potter J (1986) Molecular cloning and sequence analysis of a haploid expressed gene encoding *t* complex polypeptide 1. Cell 44:727–738

Wolfe HG, Erickson RP, Schmidt LC (1977) Effects on sperm morphology by alleles at the pink-eyed dilution locus in mice. Genetics 85:303–308

Wright WW, Parvinen M, Musto NA, Gunsalus GL, Phillips DM, Mather JP, Bardin CW (1983) Identification of stage-specific proteins synthesized by rat seminiferous tubules. Biol Reprod 29:257–270

Wyrobeck AJ, Bruce WR (1975) Chemical induction of sperm abnormalities in mice. Proc Natl Acad Sci USA 72:4425–4429

Wyrobeck AJ, Bruce WR (1978) The induction of sperm shape abnormalities in mice and humans. In: Hollaender A, DeSerres FJ (eds) Chemical mutagens: principles and methods for their detection, vol 5. Plenum, New York, pp 257–285

Yelick PC, Johnson PA, Kleene KC, Hecht NB (1985) Sequence analysis of cDNA's encoding mouse protamine 1 and 2 (MP1 and MP2) suggests that MP2 is synthesized as a precursor whereas MP1 is not. J Cell Biol 101(#5 Pt. 2):366a

Spermatogenesis in *Drosophila*

Johannes H. P. Hackstein [1]

1 Introduction

Spermatogenesis is an intensively studied developmental process in *Drosophila melanogaster* and *Drosophila hydei*. Nonetheless, some phases of the development of the germ cells are still hardly understood. The main gap of information is with respect to the germ line-soma differentiation and the formation of the gonad. Our knowledge about pole cell formation and its migration to the testis anlagen is still incomplete, and nearly nothing is known about the genetic control of these processes. In addition, some of these processes may be shared by both spermatogenesis and oogenesis. Therefore, this chapter will deal mainly with later stages of spermatogenesis, i.e., with the genetic control of the spermatogenesis sensu strictu.

1.1 A Short Description of Spermatogenesis

In both species, *D. melanogaster* and *D. hydei,* this process starts with the formation of primary spermatogonia by an unequal division of a stem cell. These stem cells occupy an apical position in the tip of the testis tube where they are anchored in the testis wall. The spermatogonium becomes invested by two cyst cells. A species-specific number of gonial divisions leads to a cyst, which contains 8 (*D. hydei*) or 16 (*D. melanogaster*) fully developed spermatogonia. All spermatogonia of one cyst are clonally derived and are interconnected by a system of cytoplasmic bridges, which resembles the ring channels of the nurse cell-oocyte complex of female *Drosophila* (Meyer 1961; Grond 1984). Development and differentiation of all cells within this syncytium are highly synchronous (Fig. 1).

The germ cells of a cyst enter meiotic prophase, becoming primary spermatocytes. These cells grow intensively and nuclear volume increases considerably. Highly specialized nuclear structures are formed during this stage in the majority of the *Drosophila* species (Meyer et al. 1961). The Y chromosome enters a period of elevated synthetic activity, forming giant lampbrush loops (Meyer 1963; Hess and Meyer 1963a, b). The primary spermatocyte stage is crucial for the further development of the germ cells. It is the longest phase of sperm differentiation, and

---

[1] Department of Genetics, Katholieke Universiteit, Toernooiveld, 6525 ED Nijmegen, The Netherlands.

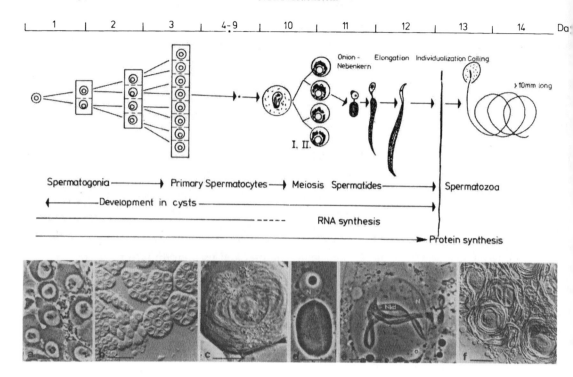

Fig. 1a–f. Schematic and photographic presentation of the spermatogenesis of *D. hydei*. Spermatogenesis in *D. melanogaster* is rather similar, but differs with respect to the number of germ cells per cyst and the time scale. The *upper scale* indicates the timing of spermatogenesis after Hennig (1967). Below the time scale, prominent stages of spermatogenesis are indicated schematically and, still lower, by photographs: **a** spermatogonia; the nucleolus associated with partially condensed chromatin floats free in the nucleoplasm. **b** Several cysts with primary spermatocytes; the nucleolus is attached to the nuclear membrane; the number of primary spermatocytes per cyst is variable. A higher magnification photograph of a primary spermatocyte is shown in Fig. 12. **c** First meiotic division; within the nucleus, residues of the lampbrush loops are visible. A layer of mitochondria (*Mi*) surrounds the nucleus. **d** Detail of a young spermatid, onion-nebenkern stage. The round nucleus contains one large refractive body, surrounded by smaller spheres. **e** Elongating spermatid with a spindlelike nucleus; the nucleus no longer contains any refractive material. **f** Coiled spermatozoa after individualization. They are still associated in bundles and their heads are attached to the head cyst cell. *N* nucleus; *NK* nebenkern; *NKd* nebenkern derivatives; *Hcc* head cyst cell; *Mi* mitochondria; *A* aster. Bars represent 10 µm in **a, c, d, e**, and 50 µm in **b** and **f; a, d, e** phase-contrast, **b, c, f** interference-contrast. Photography by W. Kühtreiber, C. Grond, and W. Hennig. (Hennig 1985a)

it is the stage with the highest RNA synthetic activity of the whole developmental pathway (Fig. 1). RNA synthesis reaches a peak in young primary spermatocytes of stage II (Hennig 1967); however, transcription ceases before the stage IV primary spermatocytes enter the first meiotic division. There is no further RNA synthesis during sperm development. Since protein synthesis continues for several days (up to the individualization stage), the messenger RNAs transcribed in the primary spermatocyte stage must be extremely stable and posttranscriptional control during these stages must be important (Olivieri and Olivieri 1965; Hennig

1967; Gould-Somero and Holland 1974). All biochemical data show that the transfer of developmental information from the genome to RNA is restricted to the primary spermatocyte stage, i.e., the diploid stage, of germ cell development. It is obvious that the full genetic complement is present only premeiotically. For example, after meiosis only half of the spermatids of a cyst contains an X or a Y, respectively. Lindsley and Grell (1969) showed in *D. melanogaster* that the complete postmeiotic development can be performed with only chromosome 4. They obtained progeny from eggs which had been fertilized by sperm which contained only chromosome 4; all other chromosomes had been removed under the influence of a meiotic mutation. Therefore, most information for spermiogenesis has to be provided premeiotically and the postmeiotic genetic content of a spermatid does not totally specify its final development.

There are, however, some exceptions to this rule: in meiotic drive systems (Peacock and Miklos 1973; Zimmering 1976) the recovery of sperm is dependent on the constitution of the haploid chromosome set. In the segregation-distortion (SD) system sperm carrying the SD gene are preferentially recovered from SD/+ males (Hartl and Hiraizumi 1976). Within one and the same cyst, where all germ cells are interconnected by cytoplasmic bridges, the non-SD-bearing spermatids degenerate (Tokuyasu et al. 1977; Hauschteck-Jungen and Hartl 1978, 1982). Therefore, postmeiotically there is a differential development of spermatids within a cyst. In this special case the genetic constitution of a spermatid is crucial for its differentiation.

Besides SD and some other meiotic drive systems a great number of meiotic mutations were identified. In the male germ line they affect not only the segregation of the chromosomes in meiosis I and II, but also control the mitotic (gonial) exchange and the mitotic segregation. A consideration of these mutations is beyond the scope of this review, since frequently these mutations are male-fertile and do not affect the development of the sperm. Meiotic mutations were reviewed by Baker and Hall (1976).

The long-lasting primary spermatocyte stage is followed by the two meiotic divisions, which are separated by a short secondary spermatocyte stage. The postmeiotic development is characterized by an enormous elongation of the spermatids that occurs without major change of the cellular volume. The nucleus elongates considerably with a simultaneous, enormous reduction of its cross-sectional area, accompanied by a compaction of the chromatin. The mitochondria fuse to form the nebenkern, which in the course of spermatid development is transformed into the paracrystalline material. The nebenkern derivative is closely associated with the flagellum which is elaborated from the centriole. After differentiation and elongation the sperm cells become individualized while their heads remain anchored in the head cyst cell. The sperm cells coil as a bundle of the 32 (*D. hydei*) or 64 (*D. melanogaster*) cells. The spiralized sperm are capable of moving and become transferred to the seminal vesicle.

Spermatogenesis is a continuous process with each developmental stage represented by a certain number of cells or cysts in the testis tube at any time after eclosion of the male fly. (However, young *D. hydei* males lack mature sperm.) Therefore, within a single testis tube one can follow the complete development of the male germ cell. The different developmental stages are found in particular zones in the testis tube, with the youngest stages near the tip of the testis and the oldest stages at the end of the testis tube, where the testis becomes the seminal vesicle (*D. melanogaster*) or the vas deferens (*D. hydei*). Spermiogenesis of *D. melanogaster* was described in detail by Tates (1971), Tokuyasu (1974a–c, 1975a, b), and Tokuyasu et al. (1972a, b).

Earlier ultrastructural studies of the differentiation of the spermatozoon (Baccetti and Bairati 1964; Bairati and Baccetti 1965; Bairati 1967; Meyer 1964, 1968; Kiefer 1966, 1970; Anderson 1967; Shoup 1967; Perotti 1969) will not be considered here since Lindsley and Tokuyasu (1980) have recently reviewed spermatogenesis in *D. melanogaster*.

Descriptions of spermatogenesis of *D. hydei* in vitro were published by Fowler (1973a); Liebrich (1981a, b); Grond (1984) and Grond et al. (1987) for development in vivo in wild-type males. Some specialized aspects, such as sperm storage and transfer, have been reviewed by Fowler (1973b).

2 The Genetic Requirements for Spermatogenesis

2.1 The Number of Genes Affecting Spermatogenesis

Spermatogenesis in *Drosophila* requires the function of a large number of genes. Natural populations harbor many male-sterile mutations in their gene pool, indicating that male-sterilizing mutations occur frequently (Lindsley and Lifschytz 1972). In mutagenization experiments a considerable number of male-sterile mutations is always recovered. Based on the extrapolated number of all possible lethal mutations on the X chromosome and the autosomes, and with the assumption that the relative mutability of male-sterile genes is a constant parameter, it is calculated that about 600 loci can mutate to male sterility in *D. melanogaster* (Lindsley and Lifschytz 1972; Lindsley and Tokuyasu 1980; Lindsley 1982). The majority of these mutants is supposed to interfere with spermatogenesis in a direct manner. Another group of mutations (temperature-sensitive lethals) act in a more indirect manner: at permissive temperature, these genes, although not lethal, affect male fertility or they render males sterile at restrictive temperature after development at permissive temperature (Shellenbarger and Cross 1977). These two authors and several other investigators observed that about 30% of the ts lethals behave as male-steriles under these conditions. This means that about 1800 additional genes can be regarded as putative male-steriles. If one assumes that these 1800 ts mutants and the 600 male-steriles belong to different collectives of genes, then 2400 genes or 36% of the genome are crucial for spermatogenesis (Lindsley 1982). This is a very high estimate. It may not be reasonable to assume that spermatogenesis needs the activity of such a high number of genes, whereas such a complex process as embryogenesis of *D. melanogaster* is governed by not more than some hundred genes (Lehmann et al. 1983; Jürgens et al. 1984; Nüsslein-Volhard et al. 1984; Wieschaus et al. 1984). Therefore, I prefer the assumption that spermatogenesis is directly controlled only by a moderate number of genes. However, spermatogenesis is extremely sensitive to metabolic stress. This leads to the high number of mutations which cause sterility by pleiotropic effects. The current dilemma is the lack of suitable screening procedures which allow the recovery of the specific male-sterile genes from the bulk of pleiotropic mutations affecting male fertility.

The Y-chromosomal, male-fertility genes are specific for spermatogenesis since the Y chromosome is completely dispensable in all somatic cells (Bridges 1916) and active only in the germ cells during the primary spermatocyte stage (Olivieri and Olivieri 1965; Hennig 1967; Ayles et al. 1973; Leoncini 1977). This completely heterochromatic chromosome (Heitz 1933) contains, in addition to a number of ribosomal cistrons (Ritossa and Spiegelman 1965), exclusively genes which are essential for male fertility and male meiosis. The lack of the Y chromosome in X/O males leads to a dramatic breakdown of spermatogenesis (see Kiefer 1966; Meyer 1968) and to a severe disturbance of male meiosis (Lifschytz and Hareven 1977; Lifschytz and Meyer 1977). Thus, the Y chromosome occupies a key function in spermatogenesis. Its genes are few but extraordinary in structure and function (see Sec. 3).

2.2 Chromosome Rearrangements Causing Male Sterility

In addition to male-fertility genes on the Y, the X, and the autosomes there are other genetic conditions affecting the course of spermatogenesis. For example, the majority of translocations between the X chromosome and the large autosomes 2 and 3 are male-sterile in *D. melanogaster* (Lindsley and Lifschytz 1972; Lindsley and Tokuyasu 1980). These T(X;A) mutants are dominant steriles, i.e., the sterility cannot be rescued by duplication of the regions around the breakpoint. For some of the translocations, it has been shown that females carrying two doses of a male-sterile X-A translocation are fertile (Lindsley and Tokuyasu 1980). However, it has been reported that X-autosome translocations of *D. melanogaster* can be fertile if fewer than ten numbered salivary gland chromosome divisions are interchanged between the X and the autosome (Lindsley 1982). It is unclear whether "numbered divisions" refer to chromosome bands and how the translocations combine the X chromosomal and autosomal parts. T(X;A) mutants can also be fertile if the breakpoints are located in the X-chromosomal centric heterochromatin (Lindsley and Lifschytz 1972).

In order to explain this chromosome sterility Lindsley and Lifschytz introduced the model of X inactivation during spermatogenesis (Lifschytz and Lindsley 1972; Lifschytz 1972). In this model, the X chromosome becomes inactivated during the primary spermatocyte stage, starting from a region in the proximal X heterochromatin. X-autosome translocations were proposed to interfere with the inactivating region, thereby disturbing the normal sequence of gene activity and inactivity during the primary spermatocyte stage; as a consequence, the process of spermatogenesis is inhibited. The cytological evidence for a precocious X inactivation in the primary spermatocyte of *D. melanogaster*, however, is poor (Cooper 1950). In *D. hydei* cytological studies with the fluorescent DNA dye DAPI reveal no evidence for X inactivation during premeiotic development (Kremer et al. 1986). In addition, Lifschytz and Hareven (1977) isolated about

600 X-chromosome male-steriles. Since some of these mutations exert detectable deviations from the normal development during the primary spermatocyte stage or during meiosis (Lifschytz and Hareven 1977; Lifschytz and Meyer 1977), these genes are likely to be active at a stage when the X chromosome is supposed to be inactive. Therefore, the reason for the sterility of X-autosome translocations remains unclear. One should perhaps consider the possibility that the translocations interfere with the dosage compensation of translocated genes. A possible imbalance of gene products, caused by an overactivity of autosomal genes close to the X breakpoint, could disturb spermatogenesis. This type of damage is also not expected to be rescued by duplications which cover the breakpoint.

3 The Y Chromosome

3.1 Functions of the Y Chromosome

For many years the heterochromatic Y chromosome was regarded as genetically inert because it carries no conventional genes detected by mutations to alleles with morphological traits. However, there is one exception which causes a visible phenotype, the bobbed locus on the short arm of the Y chromosome (Stern 1927). This locus represents the ribosomal genes, which are clustered in the nucleolus organizer region (Ritossa and Spiegelman 1965; Ritossa 1976). The bb locus is active in the somatic cells of X, bb$^-$/Y, bb$^+$ males. It, hence, escapes inactivation that might be presumed to occur by heterochromatization of the Y chromosome in somatic cells.

Another function of the Y chromosome, which is still not yet understood in molecular terms, is the modulation of position-effect variegation of genes which show this phenomenon. In most instances, extra Y chromosomes suppress the position-effect variegation of genes (Spofford 1976).

A third function of the Y chromosome, related to its control of male fertility and meiosis, lies in the collochores which are supposed to control the disjunction of the X and Y chromosome during meiosis (Cooper 1959, 1964). These pairing sites have not been characterized at the molecular or genetic level. Some observations (see below) have raised doubts on the existence of these structures.

The main function of the Y chromosome, however, lies in the control of male fertility and male meiosis. The genetic role of the Y chromosome has been reviewed earlier by Williamson (1976) and by Lindsley and Tokuyasu (1980).

3.2 The Genetics of the Y Chromosome of *D. melanogaster*

In 1916 Bridges found some exceptional, patroclinous sons in certain crosses; these males were phenotypically entirely normal, except that they were sterile. Bridges demonstrated that these patroclinous males were the products of primary nondisjunction and had an X/O constitution. Safir (1920) confirmed these observations, and so it became evident that the lack of the Y chromosome was responsible for sterility of X/O males. Stern (1929) was the first to demonstrate that both arms of the Y chromosome were needed for fertility. Based on the observation that XY^L/O and XY^S/O males, i.e., males with X-Y translocations carrying either the long or the short arm of the Y chromosome, are sterile, he postulated that each arm of the Y chromosome carries a complex (Kl or Ks) of fertility genes. He was able to show that duplication of one complex cannot compensate for the deletion of the other.

In 1939, Neuhaus attempted the genetic analysis of the fertility complexes. He had to face several problems which are still inherent to the genetic analysis of the Y chromosome. First, the Y chromosome carries no visible marker genes. Even today, there are only a few Y chromosomes available with visible marker genes; with one exception, they carry the morphological markers at the tips of the chromosome arms. Second, normally there is only one Y chromosome in males. However, two Y chromosomes or at least two Y chromosome fragments are needed for complementation analysis. Third, there is no regular meiotic crossing over between Y chromosomes or between the Y and X chromosomes. This means that different mutations can be localized only by deletion/complementation mapping. It is not possible to localize Y-chromosome mutations by genetic means because of the lack of marker genes. Only cytogenetic methods allow more precise localization of the Y chromosomal genes (see Sects. 3.3 and 3.5).

Therefore, Neuhaus tried to induce Y-chromosome, male-sterile mutations by X-ray irradiation-induced breakage of unmarked Y chromosomes. Broken Y fragments were suspected to carry a sterile mutation and the Y fragments were recovered as Y-IV translocations. These translocations were recognized by the Y-induced position effect of the fourth chromosome marker gene "cubitus interruptus" (ci). Males with a ci phenotype were tested for a patroclinous transmission of ey^+ ("eyeless," another fourth chromosome marker gene). A ci phenotype and the patroclinous transmission of ey^+ was regarded as a proof for a Y-IV translocation. However, since the Y chromosome was unmarked it was not possible to identify the fragments of the broken Y chromosome. X-rayed Bx males were crossed to XY^S; ey ci or to XY^L; ey ci females. The ey^+ ci mutants which were recovered in the F_1 males were supposed to be the result of a position effect of ci caused by an X-ray induced Y-IV translocation. The sterility which is caused by the Y break is complemented by the XY^L or XY^S received from the mother.

Neuhaus made complementation tests with the different sterile Y-IV translocations. He found four male-sterile loci on the long arm, and five such loci on the short arm of the Y chromosome. The latter number was not correct (see below). It may be that he found the correct number of four on the long arm by chance, since the method of complementation was rather complicated and suffered from the lack of a marked Y chromosome. The complementation test depended on rare nondisjunctional events. Because the Y fragments were unmarked, these exceptions could not be discriminated morphologically from their regular sibs. Therefore, large numbers of flies had to be tested for complementation.

Neuhaus also studied his Y-IV translocation in oogonial metaphase plates. Since he found no visible translocations, he concluded that the male-fertility factors were clustered at the tip of the Y chromosome, and that Y-IV translocations of the size of a normal chromosome IV could carry all the fertility genes from one arm of the Y chromosome. Since this is clearly not the case (see below), the whole publication has to be treated with caution. Nevertheless, Neuhaus designed the first rationale to dissect the complexes of male-fertility genes on both arms of the Y chromosome.

In 1960, Brosseau undertook a new attempt to determine the number of male-fertility genes on the Y chromosome. He induced male-sterile mutations on free Y chromosomes; the different sterile Y chromosomes (Y^i, Y^j) were tested for complementation in an $X/Y^i/Y^j$ constitution. In an additional experiment, the sterile Y chromosomes were complemented by different X-Y translocations.

Brosseau irradiated y/sc^8.Y, y$^+$ males with X-rays; the sc^8.Y chromosome is also known as y$^+$.Y. This Y chromosome carries at the tip of the long arm of the Y chromosome a small piece of X-chromosome material, including the y$^+$ allele. The irradiated males were crossed to homozygous XY, In(1) EN, y B females. In single-pair matings the F_1 males (constitution XY, y B/sc^8.Y, y$^+$) were crossed with XY, y B/In(1) dl-49 y w m virgins. Every F_2 was expected to give rise to two kinds of males, XY,y B/sc^8.Y*,y$^+$ males and In (1) dl-49, y w m/sc^8.Y*,y$^+$ males. If a Y-chromosome, male sterile mutation had been induced in the sc^8.Y*,y$^+$ chromosome, then the In(1)dl-49, y w m/sc^8.Y*,y$^+$ males were sterile. The mutated Y chromosome could be recovered via the XY, y B/sc^8.Y*,y$^+$ sibs. In these males every recessive male-sterile mutation on the sc^8.Y*,y$^+$ chromosome is complemented by the XY,y B compound. Consequently, these males were fertile. In the case of sterility of the In(1)dl-49, y w m/sc^8.Y*,y$^+$ males in the F_3 no y w m, non-B females would be recovered. Therefore, the F_3 vials were screened for the absence of these females. Cultures with sterile males were retested for sterility and regular segregating of the y$^+$ from the XY,y B compound. The regular segregation cultures were tested for the absence of autosome translocations by crosses to In(2L+2R)Cy for chromosome II, to In(3LR)DcxF for chromosome III, and to sv^n for the fourth. The regular segregating and translocation-free chromosomes were crossed to XY^L and XY^S chromosomes in order to determine the location of the male-sterile mutations on the two chromosome arms.

The screening of 1650 sc^8.Y chromosomes resulted in the recovery of 37 male-steriles on Y^L and 13 male-steriles on Y^S [3% ms(Y)'s]. Chromosomes with different male-sterile mutations on the same chromosome arm were tested for complementation in every possible pairwise combination by crossing the XY,y B/sc^8.Y*,y$^+$ males to y v / y v /Y:bw$^+$;bw females. From every combination 80–100 males of the X/ms(Y)^i/ms(Y)^j constitution were tested for complementation, i.e., their ability to fertilize females and thus to produce progeny.

The tests of 11 ms(Y)'s sterile in Y^S and 30 ms(Y)'s sterile in Y^L revealed two complementation groups in Y^S and 5 in Y^L. They were called ks-1 and ks-2 and kl-1 to kl-5 (Fig. 2). Frequently, several adjacent complementation groups were mutated to sterility. Brosseau was able to demonstrate that several different X-Y translocations complement single ms(Y) mutations or groups of adjacent ms(Y) mutations. The X-Y translocations had been induced by Parker and Hammond (1958) by the detachment of $\overline{XX}$ chromosomes. These translocations in combination with the ms(Y)'s permitted confirmation of the linear order of the complementation groups on the Y chromosome. Figure 2 shows that kl-4 is only characterized by two chromosomes which are kl-4$^-$ kl-5$^-$ or kl-3$^-$ kl-4$^-$, respectively. This complementation group kl-4 could not be confirmed by several authors (see below); therefore, the chromosomes kl-4$^-$ kl-5$^-$ and kl-3$^-$ kl-4$^-$ failed to complement for reasons that are unknown.

Brosseau's genetic map of the Y chromosome has been confirmed with the aid of EMS-induced mutants by Williamson (1970, 1972). Williamson isolated 96 (96 of 4236, frequency 2.3%) ms(Y)'s and 89 mutants were tested for complementa-

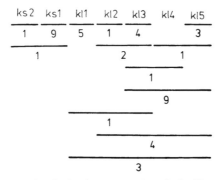

Fig. 2. Complementation map for the Y-chromosomal, male-fertility genes of *D. melanogaster*. kl-1 to kl-5 and ks-1 and ks-2 indicate the five different male-fertility genes on the long arm and the two genes on the short arm of the Y chromosome, respectively. Each *bar* characterizes a distinct type of mutation; nonoverlapping *bars* indicate complementation of the corresponding mutants. The numbers below the *bars* indicate the frequency of each mutation; a total of 45 different mutations were analyzed. kl-4 is characterized only by two overlapping mutations. Note the high frequency of mutations affecting two or more sites. (After Brosseau 1960)

tion with representatives of Brosseau's stocks. He found male-sterile mutants for each of the kl-1 to kl-5 and ks-1 and ks-2 fertility genes which had been described by Brosseau. Many of his mutants were deficient for more than one fertility factor. Unfortunately, the documentation (Williamson 1970) does not allow the identification of these stocks and provides no data for the constitution and the frequency of the male-steriles which affect several fertility genes. The mutants within every one of Brosseau's fertility factors were tested for complementation in an $X/Y^i/Y^j$ constitution (Williamson 1972).

The main result of these complementation tests was the demonstration of extensive intragenic complementation within every fertility factor; up to five complementation groups per fertility factor (kl-5) were indicated.

The reasons for this intragenic complementation remain unclear. Williamson used the double-marked $B^S Y y^+$ chromosome which allows the phenotypical control of an X/Y/Y constitution. Therefore, in contrast to earlier investigations, the genetic constitution of the tested males could be controlled unequivocally. The various Y chromosomes were classified with the aid of Brosseau's stocks. Since recently considerable genetic breakdown and mislabeling of Brosseau's stocks has been demonstrated (see Sect. 3.3), it remains unclear whether Williamson could really determine the proper fertility factors that were affected in his ms(Y) chromosomes. It is possible that his samples of ms(Y)'s were erroneously classified which might explain unexpected complementation. The evidence for the intragenic complementation as demonstrated by Williamson remains an enigma, because it was not possible to repeat these experiments (see below and Sect. 3.3).

There are two remarkable features of Williamson's work: the high number of complementation groups inferred, and the screening procedure. Williamson mutagenized y cv v f /$B^S Y y^+$ males. The Y chromosome was doubly marked by B^S and y^+ which reside in small X chromosome duplications at the tip of Y^L or Y^S, respectively (Brosseau 1958; Brosseau and Lindsley 1958). The Y chromosome was completely normal with respect to the heterochromatin markers and the male-fertility factors. The mutagenized $B^SY^*y^+$ chromosomes were recovered in F_1 l(1)J1 y w $Y^L.Y^S/B^SY^*y^+$ males. These males were mated in single crosses with C(1)TA,v/O females. This attached X chromosome is a tandem acrocentric compound which generates single, rod-X chromosomes by intrachromosomal crossing over (Merriam 1968). Because of the lethality of the l(1)J1 y w $Y^L.Y^S$/O males, the rod-X chromosomes which were generated by crossing over were found in the only class of viable males (v/$B^SY^*y^+$) that could be tested for fertility. The treated $B^SY^*y^+$ chromosome was recovered in C(1)TA,v/$B^SY^*y^+$ females. Therefore, the ms(Y)'s could be screened in the F_2 directly for sterility of the males.

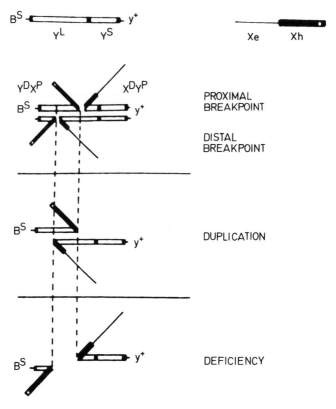

Fig. 3. Radiation-induced breaks in the Y chromosome (marked with B^S and y^+) and the X heterochromatin (*Xh, first row*) can recombine to a reciprocal X-Y translocation. Two reciprocal X-Y translocations with different breakpoints in Y^L are shown in the *second row*. The Y-distal, X-proximal (Y^DX^P) and X-distal, Y-proximal (X^DY^P) elements can be discriminated by their Y-terminal markers B^S and y^+. Males carrying only a reciprocal translocation and no free Y chromosome are fertile if the Y-chromosome breakpoint does not inactivate a male-fertility factor. The combination of the Y^DX^P and X^DY^P elements from two different "fertile" translocations leads to a duplication (*third row*) or a deficiency (*fourth row*) of Y-chromosome material between the two Y-chromosome breakpoints. If the deficiency deletes a male-fertility gene, the males carrying the two half-translocations are sterile; males with the corresponding duplication of a male-fertility gene are supposed to be fertile. The *vertical broken lines* indicate the homologous positions of both Y-chromosome breakpoints. Y^L Long arm of the Y chromosome; Y^S short arm of the Y chromosome; B^S Bar of Stone, a dominant mutation affecting the shape and the size of the eye; y^+ wild-type allele of yellow (y) body color, y^+/y or $y^+/y/y$ individuals exhibit a wild-type body color. (Kennison 1981)

In 1981 Kennison introduced the method of segmental aneuploidy (Lindsley et al. 1972) for the study of the Y-chromosomal fertility genes. This method (Figs. 3 and 4) allowed the dissection of the Y-chromosome organization with the aid of synthetic segmental deficiencies, which remove adjacent regions of the chromosome. These deficiencies were generated by the combination of the X^DY^P and X^PY^D fragments of fertile $X^h;Y$ translocations. "Fertile T($X^h;Y$)" means that males carrying these translocations were fertile in the absence of an extra Y chromosome, i.e., the translocation breakpoint did not inactivate a fertility factor.

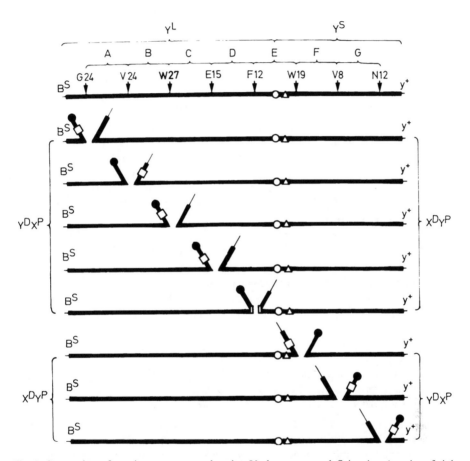

Fig. 4. Generation of contiguous, nonoverlapping Y-chromosome deficiencies. A series of eight "fertile" Y-chromosome breakpoints of T(X;Y)s (*G24, V24, W27, E15, F12, W19, V8,* and *N12*) are indicated on a schematically drawn B^SYy^+ Y chromosome. The Y-chromosome segments *A, B, C, D, F,* and *G* harbor the male-fertility genes kl-5, kl-3, kl-2, kl-1, ks-1, and ks-2. Segment *E* is devoid of male-fertility genes. A given Y-chromosome segment can be deleted by combining the left-hand element (marked with B^S) of one translocation with the right-hand element (marked with y^+) of the translocation directly below it in the diagram. The *horizontal lines* represent the Y and the *diagonal lines* the X chromosome; *heavy lines* = heterochromatin; *thin lines* = euchromatin. *Open circles* = Y centromeres; *closed circles* = X centromeres; *triangles* = bb locus ("bobbed", rDNA) on the Y; *squares* = bb locus on the X. (Hardy et al. 1981)

These translocations were induced by gamma-irradiation (4000 r) of postmeiotic stages of y w f/B^{S}Yy$^+$/Y males. The males were crossed with C(1)RM,y^{2}su(w^a)w^{a}bb/O females and the F$_1$ progeny of the genotype y w f*/B^{S}Y*y$^+$ were mated individually with C(1)RM/Y females. The F$_2$ males were screened for linkage between the treated X chromosome and the Y chromosome markers B^S, y$^+$ or both; linkage indicated a fertile translocation between X and Y. In two additional experiments Kennison also induced sterile T(X;Y)'s. In these screens irradiated males were crossed to C(1)RM/Y females instead of to C(1)RM/O females. Consequently, in the F$_1$ males the sterile Y chromosome breakpoints were complemented by an extra, fertile Y chromosome y w f*/B^{S}Y*y$^+$/Y).

For the construction of synthetic deletions in the Y^L, males possessing the translocation with the more distal breakpoint were crossed with C(1)A,y/Y^{D}X^P, B^S females, with a more proximal breakpoint. The combination of the X^{D}Y^P from the distal breakpoint with the Y^{D}X^P from the more proximal breakpoint removed all the genetic material between both breakpoints (Fig. 3). This deficiency could delete one or more fertility factor and thereby lead to sterility of the males. The reciprocal cross generated a duplication which was assumed to be fertile. The technique of segmental aneuploidy is based on two assumptions that must be realized. First, the concomitant duplications and deficiencies of the centric X heterochromatin must not sterilize the males. This assumption is apparently fulfilled since males with duplications of all the X heterochromatin are fertile; males deficient for about 80% of the X heterochromatin are also fertile (Lindsley and Sandler 1958). Second, duplications of Y chromosome material are fertile. This assumption seemed also to be fulfilled, because X/Y/Y males carrying two doses of B^{S}Yy$^+$ are fertile. There are, however, exceptions in which two doses of a Y chromosome are sterile (Grell 1969).

Thus, the analysis of 23 fertile T(X^h;Y^S) broken in Y^S and of 57 fertile translocations with breakpoints in Y^L in more than 2700 different segmental aneuploid genotypes revealed the existence of four regions in the long arm and two regions in the short arm, which are necessary for male fertility (Fig. 4). These results limited the number of the fertility genes, but they did not exclude the existence of more than one fertility factor per segment. Tests with 11 sterile translocations broken in Y^S and 30 sterile translocations in Y^L, however, did not provide any evidence for the existence of more than one male-fertility gene per segment. Thus, there are only six fertility factors on the whole Y chromosome. The tips of the chromosome arms are devoid of fertility genes, as are the regions adjacent to the Y centromere. This region can be deleted in X/Y^{LD}X^P/Y^{SD}X^P males; these males lack the Y kinetochore and are fertile. The identified fertility genes were called kl-1, kl-2, kl-3, kl-5 and ks-1 and ks-2; the most proximal ones are kl-1 and ks-1, the most distal one in Y^L is kl-5, in Y^S ks-2. It remains unclear how the relation with Brosseau's kl and ks male-fertility factors had been established, and why Kennison chose to correlate the nonexistence of one of Brosseau's complementation groups with kl-4, since no complementation tests with Brosseau's stocks had been performed. As mentioned before, kl-4 was defined only by two stocks which bore an overlapping mutation (kl-5$^-$ kl-4$^-$ and kl-4$^-$ kl-3$^-$). Williamson (1970, 1972) found three kl-4 mutations and 11 chromosomes which were triple mutants by complementation tests with Brosseau's stocks. These triple mutants were most likely kl-3$^-$ kl-4$^-$ kl-5$^-$; this was, however, not documented. The existence of kl-4 had not been questioned by Brosseau (1964), Parker (1967) and Andrews and

Williamson (1975). These authors tested X-ray-induced X-Y translocations for their ability to complement Brosseau's stocks; Parker (1967) even suggested an additional complementation group distal to kl-4. In contrast, Lucchesi (1965) also tested X-Y translocations and concluded from the lack of reciprocal exchanges between kl-3/kl-4 and kl-4/kl-5, that there was no evidence for the existence of kl-4. In addition, Hazelrigg et al. (1982); Gatti and Pimpinelli (1983) and Kennison (1983) also failed to find evidence for a fifth complementation group on the long arm of the Y chromosome. Therefore, all recent investigations argue against the existence of kl-4.

In contrast to the findings of Williamson (1972), these authors also did not find complementation between chromosomes that are sterile in one and the same fertility gene. In order to reinvestigate this phenomenon, Kennison (1983) analyzed 41 γ-ray-induced and 26 EMS-induced mutants (Fig. 5). After it had been shown that these mutations were not engaged in Y-autosome translocations, the sterile loci were determined by crosses to diagnostic synthetic deficiencies. Stocks that carried mutations in the same male-fertility gene failed to complement. Therefore, the results of Williamson (1972) could not be repeated with these samples of radiation or EMS-induced male-steriles.

Kennison (1983) also tested the old hypothesis (Neuhaus 1939) that there are male-fertility genes on the Y chromosome which escape detection because they have alleles in the X heterochromatin. The treated $sc^{7*}/B^S Y^* y^+$ chromosomes were recovered in $sc^{7*}/In(1)B^{M1}$ and $sc^{7*}/sc^{4L}sc^{8R}$ females (their sc^{7*} sons were screened for X-chromosome lethality and fertility) and in $In(1)B^{M1}/sc^{4L}sc^{8R}/B^S Y^* y^+$ females. The latter were crossed to XY/O males and the three different genotypes of sons were tested for Y-chromosome fertility. $XY/B^S Y^* y^+$ would be sterile only if a dominant, male-sterile mutation had been induced; all other ms(Y)'s would be complemented and consequently fertile. The $In(1)B^{M1}/B^S Y^* y^+$ males contained a "normal" X chromosome and would be sterile if an ms(Y) has been induced. Finally, the $sc^{4L}sc^{8R}/B^S Y^* y^+$ males should indicate a male-sterile mutation with alleles in the X heterochromatin, if they were sterile when the $In(1)B^{M1}/B^S Y^* y^+$ males with the same Y^* chromosome were fertile. The $sc^{4L}sc^{8R}$ inversion removes about 80% of the X heterochromatin. Sterility of the $sc^{4L}sc^{8R}/B^S Y^* y^+$ males also indicated the presence of fertile Y-autosome translocations (Lindsley et al. 1979). With the aid of these males Kennison detected ten $B^S Y^* y^+$ chromosomes (without Y-autosome translocations) which were sterile with $sc^{4L}sc^{8R}$ for two generations. These ten chromosomes failed to sterilize $sc^{4L}sc^{8R}$ males when retested several generations later. This could mean that these mutations were unstable and reverted to fertility within some generations. Therefore, there is no evidence for stable, Y-chromosome, male-sterility mutations which have alleles in the X heterochromatin deleted in the $sc^{4L}sc^{8R}$ chromosome.

The comparison of fertility and sterility of the three genotypes $XY/B^S Y^* y^+$, $In(1)B^{M1}/B^S Y^* y^+$, and $sc^{4L}sc^{8R}/B^S Y^* y^+$ reveals that γ-irradiation induces a large number of Y-autosome translocations which sterilize the $XY/B^S Y^* y^+$ males. These Y-autosome translocations sterilize males in a dominant manner since the XY compound provides a complete Y chromosome for complementation of all recessive Y-chromosome male-steriles. Therefore, not only X-autosome (see Sect. 2.2), but also Y-autosome translocations can behave as dominant male-steriles.

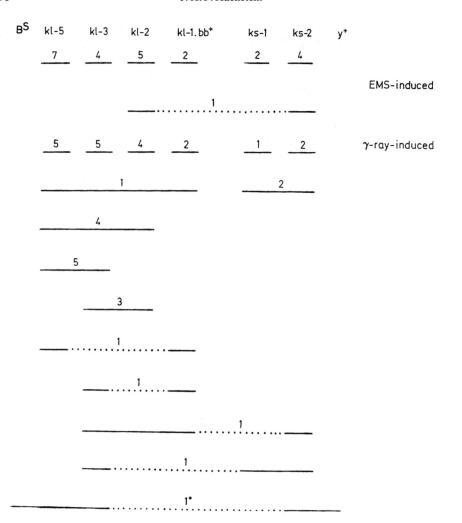

Fig. 5. Complementation map of the Y-chromosomal, male-fertility genes of *D. melanogaster*. *kl-5, kl-3, kl-2,* and *kl-1* indicate the complementation groups on the long arm, *ks-1* and *ks-2* those on the short arm of the Y chromosome. Only male-sterile Y chromosomes that are not involved in translocations are shown. *Solid lines* indicate the defective fertility regions; the *numbers* above each bar are the numbers of the mutant chromosomes of that type. A *dotted line* indicates the presence of functioning fertility regions between nonadjacent defective regions in the same mutant chromosome. The 25 mutant chromosomes in the upper half were EMS-induced, whereas the group of 39 mutants in the lower half were γ-ray induced. The chromosome indiated an *asterisk* is a ring that has lost the terminal B^S and y^+ markers. Compare with Fig. 2. (Kennison 1983)

These translocations sterilize males in a chromosomal, rather than genic, manner. In contrast, X-Y translocations can be complemented by Y-chromosome material; here, most likely the Y-chromosome breakpoint inactivates a fertility gene (Kennison 1983; Nicoletti and Lindsley 1960).

Another chromosome type of sterility is caused by three doses of the Y chromosome (Cooper 1956). Williamson and Meidinger (1979) showed that the triplication of the long arm of the Y chromosome causes the sterility. Kennison (1981) was able to demonstrate that a triplication of kl-3 is sufficient to render the male sterile.

In conclusion, the Y chromosome of *D. melanogaster* harbors only six male-fertility genes, four in the long arm, and two in the short arm. There is only one complementation group per fertility gene and there is no indication for stable male-fertility genes on the Y chromosome with alleles in the X-chromosome heterochromatin. The distal regions of both arms of the Y chromosome and the region around the kinetochore are devoid of fertility genes.

All recent investigations show that male-fertility factors can be regarded as conventional genes which mutate to their male-sterile alleles with a high frequency after irradiation or EMS mutagenization. The complementation tests and the localization of the fertility genes suffer from the lack of (interstitial) marker genes and the absence of crossing over. Nevertheless, it becomes evident that the fertility genes are distributed over large areas of both arms of the Y chromosome. However, some regions of the Y chromosome are devoid of male-fertility genes. Except for the problems that arise by the (unexplained) irregular complementation of certain Y-autosome translocations (Gatti and Pimpinelli 1983) and the dominant sterility of other Y-autosome translocations, the genetics of the Y chromosome provide no exceptional features.

It should, however, be kept in mind that all screening methods depend on complete sterility of a male caused by the mutation of a male-fertility gene. So it is not surprising that all investigators (Neuhaus, Brosseau, Williamson, Kennison, Gatti and Pimpinelli, Hazelrigg et al.) failed to find that certain regions of the Y chromosome, which do not sterilize the male, influence the male germ line. An example is the region on Y^L that controls expression of the X-linked Stellate locus. The function of the Stellate (Ste)-related sequences on the Y chromosome which interfere with the male meiosis will be described later (see Sect. 3.7).

3.3 The Cytogenetics of the Y Chromosome of *D. melanogaster*

The Y chromosome does not undergo regular meiotic recombination; in the *Drosophila* male meiotic recombination is absent. Normally, there is only the heterologous X as a partner for the Y chromosome in the heterogametic sex. Even if placed in a female there is no regular crossing over between X and Y or between two Y chromosomes; thus, it is impossible to localize the male-fertility factors by conventional mapping. There are, however, exchanges between X and Y (Neuhaus 1936, 1937; Lindsley 1955); they occur rarely and their nature is unknown. Use of these X-Y exchanges for the localization of the male-fertility genes has never been attempted except in a single experiment published by Neuhaus (1939).

The heterochromatic Y chromosome is underreplicated in salivary gland nuclei and is included in the chromocenter with other heterochromatic constituents of the genome. Thus, a cytogenetic analysis of the Y chromosome in salivary gland preparations is not possible. The only other reasonable cytogenetic approach is the analysis of metaphase Y chromosomes from diploid cells, usually larval neuroblast cells.

Cooper (1959) analyzed aceto-orcein stained metaphases and was able to demonstrate that the Y chromosome can be subdivided into seven heavily staining blocks which are separated by more or less prominent constrictions. A more detailed cytogenetic analysis became possible with the fluorescent dye Hoechst 33258 (Holmquist 1975). Holmquist was able to distinguish 11 blocks of bright fluorescent, dull fluorescent, and nonfluorescent regions in metaphase Y chromosomes. Gatti et al. (1976) improved the resolution of the Hoechst 33258 staining. They also used quinacrine staining, sometimes applied sequentially with the Hoechst 33258 staining, in order to analyze the fluorescence banding of prometaphase Y chromosomes. Pimpinelli et al. (1976) additionally applied C- and N-banding. If the same chromosome was sequentially stained with quinacrine, Hoechst 33258, and N-banding, the resolution of the banding methods could be improved in such a way that 25 regions of the prometaphase Y chromosome could be discriminated. This resolution should be sufficient to localize the male-fertility factors (Fig. 6).

Kennison (1981) was the first to publish a drawing based on Gatti's map of the Y chromosome. In this drawing the breakpoints of a number of sterile and fertile T(X;T)translocations were indicated. A genetic map of the Y chromosome constructed from the breakpoint frequencies suggested that there were four regions on the long arm and two regions on the short arm which could be correlated with the male-fertility factors. Unfortunately, it is impossible to identify the trans-

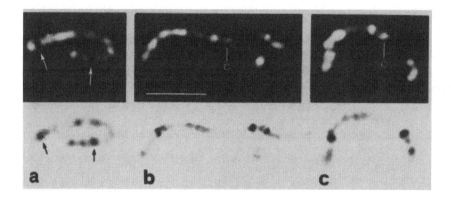

Fig. 6a–c. Y chromosomes of *D. melanogaster* sequentially stained with Hoechst 33258 and N-banding. The *upper row* shows the Y chromosomes in fluorescence microscopy (Hoechst stained); the *lower row* shows the same chromosomes in phase-contrast after N-banding. Note that the N-bands correspond to the nonfluorescent regions of the Hoechst-stained chromosomes (*arrows*). **a** Oregon R (wild-type); **b** and **c** B^{S}Yy$^+$ Y chromosomes. The wild-type and the B^{S}Yy$^+$ Y chromosomes differ only in the presence of extra fluorescent regions at the chromosome ends; *c* centromere. *Bar* represents 5 μm. (Gatti and Pimpinelli 1983)

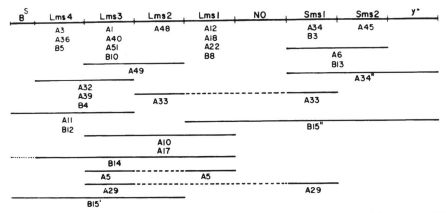

Fig. 7. Complementation map of the male-sterile mutations recovered by Hazelrigg et al. (1982). *Lms1* to *Lms4* represent the four complementation groups on the long arm. *Sms1* and *Sms2* represent the two complementation groups on the short arm of the Y chromosome. Most likely these complementation groups indicate the male-fertility genes kl-1, kl-2, kl-3, and kl-5 on the long arm and ks-1 and ks-2 on the short arm. (However, this has to be confirmed by test crosses to Brosseaus's stocks.) The *letter code* below the complementation groups identifies the different mutations. *Solid lines* indicate regions which fail to complement. The *dotted lines* indicate the presence of functional fertility regions between nonadjacent defective regions in the same mutant chromosome. *NO* nucleolus organizer; B^S and y^+ marker genes at the tips of the Y chromosome. (Hazelrigg et al. 1982)

locations by their stock numbers, and the reliability of the method was not documented by photographs.

Hazelrigg et al. (1982) obtained 29 X-ray-induced male-steriles from screening 1020 B^SYy^+ chromosomes. Complementation tests of all 29 stocks inter se revealed the existence of four complementation groups on the long arm and of two on the short arm (Fig. 7). Staining with Hoechst 33258 showed that many of the sterile Y chromosomes carried deletions and inversions. Hazelrigg et al. were able to discriminate in the Y chromosome ten bright and dull fluorescent knobs, which were separated by constrictions. Figure 8 indicates the deleted material and the inversions that could be identified. This led to the cytogenetic map of the Y chromosome shown in Fig. 9.

The analysis of a number of Brosseau's tester stocks demonstrated not only that mislabeling had occurred with many of the stocks, but also that the stocks had undergone a considerable genetic breakdown during the 2 decades of their existence (Hazelrigg et al. 1982). Many stocks were sterile in other complementation groups than indicated on the label; in many cases they were sterile in more than one fertility gene. Hazelrigg et al. showed also that in their newly induced stocks genetic changes took place. This was easily detected by the concomitant loss of the markers B^S and/or y^+. They concluded that this breakdown of the chromosomes might be caused by the gonial "exchanges" which occur in *Drosophila* males with a frequency of less than 0.1% (Neuhaus 1937; Lindsley 1955). This "exchange" is, however, not comparable to a regular crossing over, since the

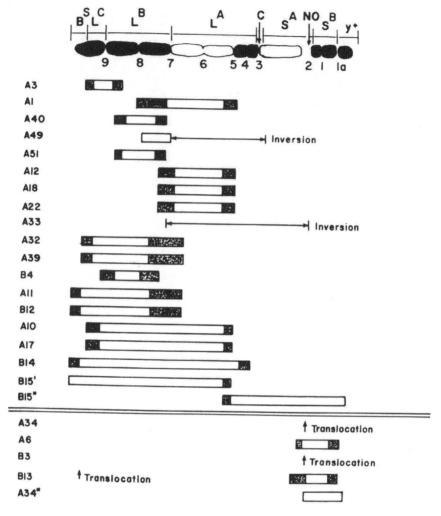

Fig. 8. A summary of the cytological examination of male-sterile Y chromosomes recovered by Hazelrigg et al. (1982). At the *top*, a drawing of a Hoechst 33258 stained $B^S Y y^+$ chromosome from a larval neuroblast metaphase plate. The *black dots* indicate Hoechst – bright regions; *white parts* of the drawing indicate dull fluorescent regions of the Y chromosome. The L^A, L^B, L^C, S^A, $S^{B,}$ and *NO* (nucleolus organizer) designations correlate these regions with the map of Cooper (1959). The constrictions between the knobs are numbered. *C* indicates the centromere; B^S and y^+ the (X chromosome-derived) marker genes. The *open bars* represent the extent of the deleted material in the different mutant chromosomes. The *dotted regions* indicate the uncertainty as to the endpoints of the deletions. (Hazelrigg et al. 1982)

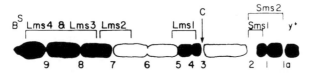

Fig. 9. Localization of the Y-chromosome, male-fertility factors on a Hoechst 33258 stained metaphase chromosome from larval neuroblasts from *D. melanogaster*. *Lms1* to *Lms4* (corresponding to kl-1, kl-2, kl-3, and kl-5) indicate the fertility factors on the long arm, *Sms1* and *Sms2* (ks-1 and ks-2) those on the short arm. For further information see Fig. 8.
(Hazelrigg et al. 1982)

introduction of a ring XY compound chromosome did not inhibit the breakdown of the Y chromosomes.

In 1983, Gatti and Pimpinelli published an extended genetic and cytogenetic analysis of 206 Y-autosome translocations and 24 sterile sc^8.Y chromosomes. The Y-autosome translocations had been induced by Lindsley et al. (1972) for the study of segmental aneuploids; the $sc.^8Y$ chromosomes were from Brosseau (1960) and from other sources. Because complementation tests with different T(Y;A)'s inter se revealed irregular complementation behavior, the T(Y;A)'s were tested with different ms(Y)'s. In this combination, the complementation was regular. A detailed complementation analysis showed that four fertility factors were present on the long arm and two on the short arm. A careful cytogenetic analysis, based on the subsequent staining with quinacrine, Hoechst 33258, and N-banding, made it possible to localize the fertile and sterile breakpoints precisely (Fig. 10). It is obvious that the distribution of the breakpoints is nonrandom: breaks frequently occurred in the Xh blocks which harbor the B^S and y^+ markers. Furthermore, the nonfluorescent, N-banding blocks 3, 21, 25D, 23, and 14 are broken preferentially. Male-sterilizing breakpoints are located preferentially in the nonfluorescent N-bands (3, 10, 14, 21, 23, 25D). A more detailed representation of the breakpoints and the deletions of the different ms(Y) stocks allowed a precise localization of the fertility genes (Fig. 11).

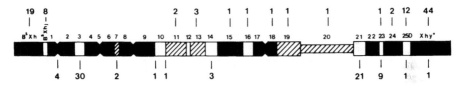

Fig. 10. Schematic representation of a Hoechst 33258 stained B^SYy^+ chromosome from larval neuroblasts as seen by Gatti and Pimpinelli (1983). The authors discriminate, in addition to the terminal blocks carrying the B^S and y^+ markers, 25 blocks of heterochromatin (*small figures at the top* of the drawing). *Black areas* represent bright fluorescent regions; *hatched areas* dull regions; *open areas* nonfluorescent regions of the chromosome. The region h20 corresponds to the nucleolus organizer, the constriction between h17 and h18 to the kinetochore. The *large figures above* the drawing indicate the frequency and localization of the "fertile" Y-chromosome breakpoints of the T(Y;2)s and T(Y;3)s; the *figure below* the drawing show the "sterile" breakpoints of the T(Y;2)s and T(Y;3)s analyzed. The term "fertile" breakpoints means that males with such a T(Y;A) are fertile in the absence of an extra Y chromosome; "sterile" indicates the sterility of such males. Compare to Figs. 8 and 9. (Gatti and Pimpinelli 1983)

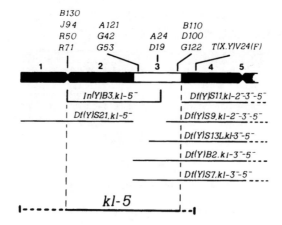
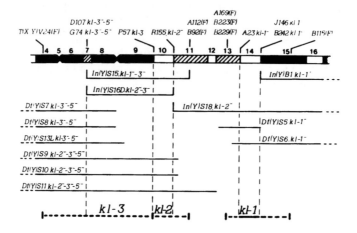
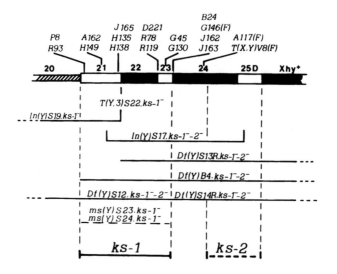

The kl-5 locus is defined by a series of 12 sterile breakpoints in h1/h2 and h3. The next fertile breakpoint is located in h4 [T(X;Y)V24]. Therefore, the sterile breakpoints are scattered over a region which roughly contains 8% of the Y-chromosomal DNA; they inactivate one and the same fertility factor. For kl-1 and ks-1 regions of a similar size (3%, respectively, 8% of the Y chromosome) characterize the functional domains of these fertility genes. The loci kl-3 and kl-2 may be of a similar size, but the scarcity of sterile breakpoints in these regions prevented a better characterization.

The distribution of sterilizing breakpoints from translocations which do not complement over such large areas raises the question whether the genes have gigantic dimensions or whether there are distance effects of the breakpoints which cause a kind of position effect. A the moment, the state of knowledge of *D. melanogaster* does not allow a conclusive answer. In *D. hydei*, however, it can be shown that male-fertility genes produce transcripts of a size which exceeds 1000 kb (see Sect. 3.6).

Of interest is the structural relationship of the male-fertility genes with the simple sequence DNA satellite of *Drosophila melanogaster*. Satellite sequences account for about 70% of the Y-chromosomal DNA of this species. Most of the satellites occur in blocks which have been localized on the Y chromosome by in situ hybridization (Peacock et al. 1978; Appels and Peacock 1978; Steffensen et al. 1981; Hilliker and Appels 1982). Gatti and Pimpinelli (1983) assumed that the Hoechst-bright bands contain the 1.672, 1.686, and 1.697 g/cm^3 satellite DNAs, whereas the N-bands are correlated with the distribution of the 1.705 g/cm^3 satellite DNA. However, this correlation between satellite sequences and bands of the Y chromosome needs experimental confirmation. At the moment, it cannot be excluded that satellite sequences form part of the male-fertility genes.

3.4 The Genetics of the Y Chromsome of *D. hydei*

In 1961 Meyer et al. described *"phasenspezifische Funktionsstrukturen"* in the primary spermatocyte nuclei of *D. melanogaster*. It was derived that these intranuclear structures originated from the Y chromosome, because they were absent in X/O males. Subsequently, it became evident that these structures are much more prominent in other *Drosophila* species. These structures were well differentiated in primary spermatocyte nuclei of *D. hydei* and some related species (Meyer 1963; Hess and Meyer 1963a, b). The chromosome specializations were identified as lampbrush loops formed by the Y chromosome (Fig. 12). Cytological investi-

---

Fig. 11. Cytogenetic map of the Y-chromosomal, male fertility genes. For explanation of the schematic Hoechst 33258 stained metaphase Y chromosome see Fig. 10. The *figures* and *letters* above the schematic Y chromosome indicate Y-autosome (or X-Y) translocations; the Y-chromosome breakpoints are "sterile" unless an "*F*" indicates fertility. The *lines below* the Y chromosome show the extensions of Y-chromosome rearrangements. The extensions of the different fertility genes as defined by the breakpoints of sterile, noncomplementing rearrangements are shown by *thick, solid lines*. The *broken lines* indicate the proximal and distal limits of each locus as defined by the nearest fertile breakpoints. Note the considerable extension of these areas. (Gatti and Pimpinelli 1983)

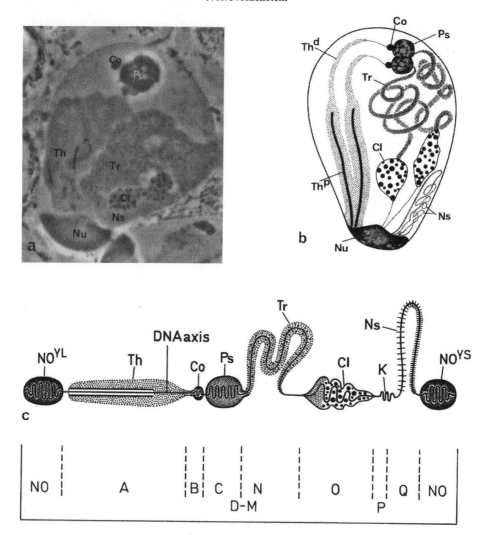

Fig. 12 a–c. Primary spermatocyte nucleus of *D. hydei*. **a** phase-contrast; **b** diagram; **c** single Y chromatid with the lampbrush loops and the genetic map of the male-fertility genes *A–Q*. **a** and **b** show the five large lampbrush loops (*Ns* nooses; *Cl* clubs; *Tr* tubular ribbons; *Ps* pseudonucleolus; *Th* threads; *Co* cones; *d* distal; *p* proximal) and the nucleolus (*Nu*, closely attached to the nuclear membrane). The loops are present as pairs, since the primary spermatocyte stage represents a meiotic prophase with a 4C DNA content. In **c** a single Y chromatid is shown schematically; it represents an open-chain model since there are no indications for a chromomeric organization of the Y chromosome in the primary spermatocyte. The correlation between the lampbrush loops and the genetic complementation groups is indicated. *Bar* represents 10 μm. (Hackstein et al. 1982)

gations proved the existence of five pairs of giant lampbrush loops, each with a well-defined morphology in *D. hydei*. Loop pairs are seen because the loops occur in the meiotic prophase which has a 4C DNA content. (For the definition and characterization of lampbrush loops, see Callan 1986.) Names of the lampbrush loops are based on their characteristic morphology: nooses (Ns), clubs (Cl), tubular ribbons (Tr), pseudonucleolus (Ps), and threads (Th) (Fig. 12). During the early years, investigations on the Y chromosome of *D. hydei* were concerned mainly with the demonstration that the lampbrush loops are ordered linearly on the Y chromosome and that all of them must be present to guarantee the fertility of the males (Hess 1965a, b, 1967a; Beermann et al. 1967). For this purpose, a series of X-Y translocations were induced by detachments of attached X chromosomes. In addition, two Y-autosome half-translocations [T(Y;A)1 and T(A;Y)2] were discovered accidentially in a new vermillion mutant. T(X;Y)/O males for the cytological inspection were generated by the induction of primary nondisjunction in females until Beck (1976) constructed an attached X chromosome which permitted the recovery of $\bar{X}\bar{X}$/O females on a large scale.

The cytology of a series of T(X;Y)/O males (metaphase plates and the morphology of the primary spermatocytes) proved the linear order of the lampbrush loops on the two chromosome arms (Hess 1965b). Two fertile mutants of the thread loops demonstrated that the lampbrush-loop structure was indeed autonomously controlled by the Y chromosome. The cytological manifestation of the loop mutations was expressed autonomously in males with two Y chromosomes (Hess 1965a). Combinations of X-Y translocations of various lengths with the two Y-autosome translocations showed that the presence of all of the five lampbrush loops was necessary for the fertility of the males. For some loops duplications of one loop did not compensate for the deficiency of another loop (Hess 1967a). However, this has not been demonstrated systematically for all loops.

In some of his drawings Hess (1967a) illustrated the existence of two pairs of nooses (Noo$_1$ and Noo$_2$) and two pairs of tubular ribbons (T$_1$, T$_2$). The drawing of two pairs of nooses was not discussed; no evidence for the existence of two pairs of nooses was presented. In our experience we have never seen two pairs of this loop (see Fig. 12), and we have no evidence for the existence of a second male-sterile complementation group on the short arm (see below). The existence of a second pair of tubular ribbons is also questionable. As an argument for a second tubular ribbon Hess used the stock 340/7 [T(X;Y)48]. This stock has modified tubular ribbons, is a rather short translocation, and is sterile in combination with T(A;Y)2 (loops Th and Ps). Therefore, Hess discussed the lack of one (or more!) extra pairs of tubular ribbons as one possibility for the sterility of T(X;Y)48/T(A;Y)2 males. The observation in young primary spermatocytes of only one unexpanded Tr loop strongly argues against the existence of two pairs of tubular ribbons. In our opinion, T(X;Y)48 is sterile with T(A;Y)2 because it has one modified (sterile?) Tr or it lacks some other male-fertility genes which do not form loops (see below).

Inactivation of one or more loops of a translocation rendered the males carrying them sterile, unless the inactivated loops were compensated by active ones (Hess 1968, 1970). Therefore, Hess concluded that it was very likely that the lampbrush loops represent male-fertility factors. This assumption was confirmed by us (Hackstein et al. 1982). In about 30 000 crosses with 125 different fertile and sterile X-Y translocations and 136 sterile Y chromosomes the genetic constitution and structure of the Y chromosome was investigated. It was shown (Leoncini 1977; Hackstein et al. 1982; Hackstein 1985) that the induction of male-sterile mutations can modify or abolish one (or more) lampbrush loop(s); a considerable fraction of the ms(Y) mutations, however, has no effect on the lampbrush-loop cytology. Within one and the same complementation group we found male-steriles with a normal cytology, with a modified morphology of one lampbrush loop, or with one loop missing. Therefore, it can be concluded that (1) a lampbrush loop is a male-fertility gene and (2) that each lampbrush loop contains only one complementation group. The existence of 16 complementation groups (A-Q) was demonstrated. We correlated complementation group A with the threads, C with the pseudonucleolus, N with the tubular ribbons, O with the clubs, and Q with the nooses. The other genes in the complementation groups (B, D, E, F, G, H, I, K, L, M, P) do not form visible loops. Complementation group B is well defined by a number of mutants, but it is not possible to correlate B with a lampbrush loop on the basis of the criteria mentioned above. Complementation groups D, E, F, were characterized only by four complementing/noncomplementing T(X;Y)/Y-fragment pairs. No ms(Y) mutations in these complementation groups were found on free Y chromosomes. Moreover, some of the Y fragments which characterized D, E, and F were unstable and became inactivated. Until the existence of the complementation groups D, E, and F can be demonstrated by mutants in free Y chromosomes, their real existence must be questioned.

The existence of the six complementation groups G to M is also problematic. These complementation groups are occupied by six temperature-sensitive male-steriles. These chromosomes had been induced by Leoncini (1977); they differed in their temperature-sensitive phases. He also presented evidence for their location in at least two different complementation groups. Evidence for the location of these mutants in different complementation groups came from their complementation in $X/Y^i/Y^j$ males. These complementation tests were, however, difficult to interpret, because two of the ts mutants were leaky. Another problem arose with the phenomenon of "synthetic sterility" which accompanied the complementation tests with ts mutants. It has been shown that two Y chromsomes, one with a lesion in complementation group A, the other in complementation group Q, do not complement in a X/Y/Y male. Another case of synthetic sterility was found in the complementation of ms(Y)Q7 with different T(X;Y)'s of increasing lengths; this mutant, sterile in Q, can be complemented with short T(X;Y)'s which carry only the complementation groups O, P, and Q. It is, however, completely sterile with translocations which are longer and which possess more complementation groups than O, P, and Q. These problems could account for the difficulties with mapping of the ts mutants mentioned above. Therefore, we cannot exclude that the six complementation groups for the ts cluster were an overestimation. There exists the possibility that the ts cluster belongs to the loop Tr, because this cluster is not complemented by T(X;Y)48 and T(X;Y)49. If the fact, that both translocations carry modified tubular ribbons, implies that both translocations are sterile in Tr (N), then the ts cluster may be correlated with complementation group N.

These considerations show that the estimate of 16 complementation groups is high. The lowest estimation of the number of complementation groups would consider only the existence of the loop-forming sites A, C, N, O, Q, and the non-loop-forming sites B and P. This estimation of seven complementation groups is closer to the six complementation groups identified on the Y chromosome of *D. melanogaster*. Thus, the genetic analysis of the Y chromosome of *D. hydei* does not reveal many differences in comparison with the Y chromosome of *D. melanogaster*. The number of complementation groups may be higher in *D. hydei*, but the principal organization of the genetic information into a small number of genes is identical.

One phenomenon of *D. hydei* is of special interest: the inactivation of Y-chromosome loci. This phenomenon is more easily detected in *D. hydei* than in *D. melanogaster*, because the appearance of cytologically detectable lampbrush loops can be taken as an indication for genetic activity. This does not mean that every time one detects a "normal" lampbrush loop, one is looking at a biologically fully functional, male-fertility gene. Genetic analysis of a great number of male-sterile Y chromosomes has shown that even a male-sterile mutation in a loop-forming site can be without any effect on the loop morphology. On the other hand, (even partial) inactivity of a loop is always accompanied by a loss of function of the corresponding male-fertility gene. Inactivation occurs in newly induced X-Y translocations (Hess 1968, 1970), but also in EMS-treated, free Y chromosomes (Hackstein et al. 1982). It appears as a mutational event, because frequently in different males of a culture bottle different stages of inactivation can be observed. Hess claimed that he had observed even within a single male different cysts with different degrees of inactivation (Hess 1970). With different degrees of inactivation is meant different degrees of loop development or different numbers of adjacent, inactivated loops. This inactivation occurs as a rule within a few generations in "hybrid" X-Y translocations where the Y chromosome is derived from *D. neohydei* and the X from *D. hydei* (Hennig I 1982)). [The Y chromosome of *D. neohydei* is very similar to that of *D. hydei*, i.e., the number and the morphology of the lampbrush loops differ only slightly; however, the localization of the loops on the Y chromosome is different (Hennig 1977; Hennig I 1978; Hennig 1985a).] An even more extreme example for this phenomenon is the regular inactivation of X-Y translocations in *D. eohydei*. If X-Y translocations are constructed by the induction of detachments of attached X chromosomes in the presence of a Y chromosome, then the recovered X-Y translocations become gradually inactivated within the first three generations after the induction (Hackstein, Cornelissen, Hennig unpublished; Fig. 13).

In *D. hydei* this process of inactivation may occur late after the induction, even after more than 20 generations. If one follows this process over a number of generations, it is striking that the inactivations occur gradually. The inactivation starts with distal loops and proceeds in a proximal direction. Similar observations have been made with the Yw^{m}Co chromosome. This chromosome carries a w$^+$ N$^+$ duplication of about 13–16 salivary gland chromosome bands; it is derived from the w^{m1} rearrangement (van Breugel 1970, 1971; van Breugel and van Zijll Langhout 1983). The Yw^{m}Co chromosome undergoes frequent exchanges with the X chromosome and other Y chromosomes (Hackstein et al. 1987b). In the

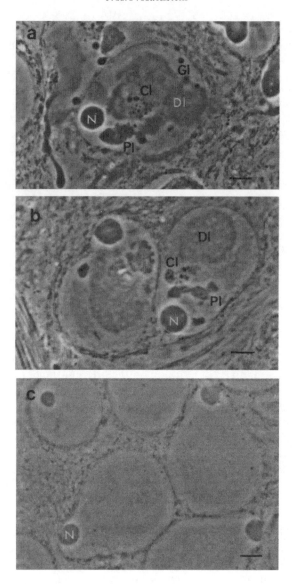

Fig. 13 a–c. Inactivation of X-Y translocations in *D. eohydei*. **a** A wild-type X/Y primary spermatocyte nucleus with the full complement of lampbrush loops. **b** Males with the X-Y translocation eo 28 developed all lampbrush loops in a T(X;Y)/O constitution (second generation after induction of the translocation). Apparently the males carried a complete X-Y compound chromosome since they became fertile and produced progeny in a cross with irradiated X/X females (irradiated with 1000 R for the induction of primary nondisjunction). Third generation patroclinous sons displayed a variable inactivation of lampbrush loops. No male developed all lampbrush loops; several males displayed only the loop Pl (not shown), other sibs no loops at all **c**. These males exhibited a typical X/O phenotype. *N* nucleolus; *Pl* pseudonucleoluslike loop; *Cl* clublike loop; *Dl* diffuse loop; *Gl* granular loop. *Bar* represents 10-μm. (Photographs by W. Hennig)

course of the transpositions of the w^mCo element, which are possibly mediated by ribosomal IVS (intervening sequences) sequences, frequently inactivation of loop-forming sites occurs. The molecular basis of these inactivations is completely unclear. The clue to this phenomenon may reside in the heterochromatic capacities of the Y chromosome (Hennig 1986); in some aspects it is reminiscent of a stable position effect (Spofford 1976) or the inactivation of the X chromosome in mammals (Lyon 1961).

The formation of lampbrush loops by active male-fertility genes provides a unique tool to screen for mutants which control the activity of the Y chromosome. Our extended genetic studies of the Y chromosome (Hackstein et al. 1982; Hackstein 1985) indicate no trans-acting Y chromosome mutations. Therefore, genes controlling the coordinate expression of the Y-chromosomal male-fertility genes should be located on the X chromosome or on the autosomes. Lifschytz (1974, 1975) recovered three X-chromosome mutants which inhibited the regular development of all (or only a part) of the lampbrush loops. The phenotype of these mutants is, however, rather unspecific: (premature) degenerating spermatocytes are found. In extended screens of autosomes 2, 3, and 4 we also recovered a number of autosomal male-steriles which interfered with the cytology of the primary spermatocyte in a rather unspecific way. Two mutations affected the Ps, probably by the lack of a protein constituent, two other mutations destroyed the structural integrity of both the Ps and the Th. One mutant on chromosome 2 specifically removes the compact, proximal part of the Th. All these mutants are clearly not regulatory mutants. However, two recessive mutations were recovered on chromosome 3 which might be regulatory mutations (Hackstein et al. 1987a). One of these mutants, ms(3)5, completely inhibits the formation of all lampbrush loops at 26 °C. At 18 °C all loops are formed, however, with a modified morphology. Temperature shifts from 18 °C to 26 °C and from 26 °C to 18 °C with mutant males indicate that the temperature-sensitive phase is in the early spermatogonium or in the stem cell. At this developmental stage apparently a decision is made, whether lampbrush loops are formed or not. This mutation clearly represents a regulatory mutant which is located at the basis of the spermatogenic pathway. It argues for a global control of all lampbrush loops. We cannot exclude, however, the possibility that the Y-chromosomal genes might be individually controlled as well. Even a single lampbrush loop, if placed in a suitable X-Y translocation, acquires its characteristic shape at the proper time in the absence of the rest of the Y chromosome (Hess 1965b, 1967a).

In addition, it is possible to obtain fertile species hybrids between *D. hydei* and *D. eohydei* (Hennig I 1978, 1982; Hennig 1977; Schäfer 1978, 1979). The literature concerning species hybrids has been reviewed recently (Hennig 1985a). It is evident that *D. hydei* offers possibilities for the study of the genetics and the control of Y-chromosomal genes, that are not easily realized in *D. melanogaster*.

3.5 The Cytogenetics of the Y Chromosome of *D. hydei*

As in *D. melanogaster*, metaphase chromosomes from brain ganglia are the only material which is suitable for cytological analysis of *D. hydei* Y chromosomes. Hess (1965b) undertook the first attempts to localize lampbrush loop-forming sites on lacto-orcein stained Y chromosomes. He compared the lengths of different translocated Y fragments with the length of a normal Y chromosome and determined the relative deletion of Y material from the translocation. This comparison enabled Hess to estimate the lengths of translocated fragments in tenths of the length of the long arm of the Y chromosome. The translocation 290/1 [T(X;Y)58] made it possible to locate the nooses on the short arm of the Y chromosome. The clubs were localized by the length of the translocation 340/2 [T(X;Y)50] in the proximal tenth of the long arm. The translocation 340/7 [T(X;Y)48] located the tubular ribbons to the proximal 2/10 or 3/10 of the long arm. The distal limitation for the pseudonucleolus is formed by translocation 340/10 [T(X;Y)20], the proximal by the Y fragment of the stock w^{m1}. In terms of the length scale this is 8/10 or 9/10. The threads are characterized by translocation 340/10 [T(X;Y)20]; they occupy the 9/10 or 10/10 segment (Hess 1965b, Fig. 18). This location of the lampbrush loops has been summarized in Fig. 1 of Hess (1967a); this figure, however, does not correctly indicate the location of Cl, Tr, and Ps as it has been described in Hess (1965b) and as it was drawn in Fig. 4 of Hess (1967a). The localization of the lampbrush loops depends on length differences between different translocations, which can be estimated only roughly. Therefore, precise localization of the loops is not possible. In addition, there is a considerable variation of the chromosome lengths between different metaphase plates and different preparations. The most important conclusion which can be made from these studies is that a part of the Y chromosome, located between 4/10 and 8/10, does not carry any lampbrush loop-forming sites.

The preparation of the metaphase plates and their lacto-orcein staining described by Hess (1965b) prevented the discrimination of structural details of the Y chromosome. In principal, this stain should allow the detection of blocks of heterochromatin as in *D. melanogaster* (Cooper 1959). Hoechst 33258 and quinacrine staining allowed the discrimination of different blocks of chromatin (Gatti et al. 1976). In contrast to *D. melanogaster*, the Y chromosome of *D. hydei* does not show any N-banding (Pimpinelli et al. 1976). Giemsa staining, TAG staining,

Fig. 14. Cytogenetic map of the Y chromosome of *D. hydei*. The chromosome diagram at the *top* shows a schematical Y chromosome from a larval neuroblast metaphase plate after staining with Hoechst 33258. The *dark areas* indicate bright fluorescent regions, the *crosshatched areas* dull fluorescent, and the *white areas* nonfluorescent regions. Region 6 represents the main constriction in Y^L, region 12 the nucleolus organizer on Y^S. C indicates the kinetochore (After Bonaccorsi et al. 1981). The block of *horizontal lines* shows the lengths of Y chromosomes which were analyzed by Hoechst 33258 staining. The *dotted lines* indicate regions of controverse or ambiguous interpretation. The *filled* and *open triangles* delimit two alternatives for the chromosome ms(Y)PE3. The *capital letters* after the stock numbers indicate the complementation groups carried by the Y chromosome. A "?" marks complementation groups that cannot be demonstrated by genetic means or that are uncertain because of inactivations in the stocks. The lengths of the different translocations allow the localization of the large lampbrush loops (*Th, Ps, Tr, Cl*). The localization of the loop *Ns* is achieved with the aid of in situ hybridization (Vogt and Hennig 1983). The localization of the nonloop-forming fertility genes is also indicated. The localization of the complementation groups *G–M* is problematic since the translocations available do not permit a precise localization (see text). (Based upon unpublished information of S. Bonaccorsi and J. Hackstein)

Spermatogenesis in *Drosophila*

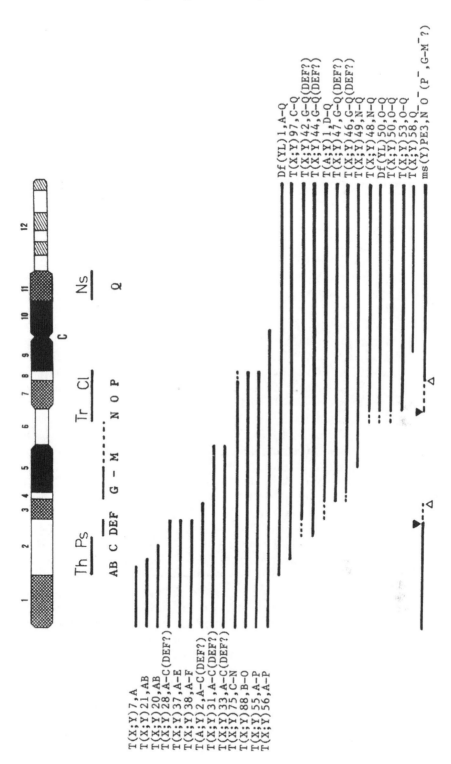

G-, and C-banding did not significantly improve the resolution of the staining (Beck and Srdić 1979). Only improvement of the Hoechst 33258 staining method increased the resolution. With a special procedure the discrimination of 12 very bright, dull, or very dull fluorescent regions becomes possible (Bonaccorsi et al. 1981). Sometimes the regions are separated by clearly discernible constrictions. The staining pattern resembles that of *D. melanogaster;* it is, however, simpler.

The analysis of Hoechst 33258 stained prometaphase Y chromosomes from 27 different X-Y and A-Y translocations and one sterile Y chromosome made it possible to correlate the complementation groups with regions of the Bonaccorsi map (Bonaccorsi and Hackstein 1987). Block 1 on the tip of the long arm is devoid of fertility genes (Fig. 14). Complementation group A (threads) is characterized by T(X;Y)7, complementation group B by T(X;Y)21; both complementation groups occupy the distal fourth of the very dull fluorescent region 2. Directly adjacent to A and B follows complementation group C (pseudonucleolus) occupying approximately one-half of block 2. The rest of block 2 may be covered by the complementation groups D, E, and F as indicated by the length of T(X;Y)38. It is possible, however, that complementation group C extends to the border of the regions 2 and 3 or even occupies a part of region 3. This is suggested by molecular studies, which make a spreading of transcribed Ps sequences into region 3 likely (Hulsebos et al. 1984; Huijser, Hennig, Hulsebos and Hackstein, unpublished results). The Y chromosome contains 42×10^3 kb, roughly 9.5% of the genomic DNA (Zacharias et al. 1982). Thus, region 2 harbors approximately 6000 kb DNA, which are distributed over the three complementation groups A, B, C (and perhaps D, E, F) with the lampbrush loops Th and Ps. The existence of very large transcripts which are derived from Th and Ps is in accordance with this calculation (see below).

The location of the complementation groups G-M, occupied by the cluster of ts mutants, is difficult to establish because we do not have translocations which separate these complementation groups from each other and from neighbouring groups. The location of G-M in blocks 4 and 5 is made likely by T(X;Y)46 and 47 which complement, and T(X;Y)48 which does not complement the ts cluster. This location implies, however, that T(X;Y)31 and 33 are partially inactivated, because they do not complement G-M. Alternatively, if T(X;Y)31 and 33 are not inactivated, the complementation groups G-M should be located close to N (Tr) in region 6. This complementation group N(Tr) is located in regions 6 and 7, because both T(X;Y)48 and T(X;Y)75 develop the lampbrush loop Tr. Complementation group O (Cl) and P are located close to N, in a more proximal location in regions 7 and 8. Again, two lampbrush loops are clustered in a large block of the Y chromosome. Regions 6, 7, and 8 occupy about 20% of the Y-chromosomal DNA; consequently, they contain more than 8000 kb DNA.

Region 9 is devoid of fertility genes, since no translocation and no sterile Y chromosomes indicate additional complementation groups on the long arm. On the short arm there are no translocations which lack complementation group Q (Ns). Therefore, Q (Ns) can be located to regions 10 distal and 11 only with the aid of in situ hybridizations of Ns-specific DNA sequences (Vogt et al. 1982; Vogt and Hennig 1983). Region 12 contains the nucleolus organizer of the short arm (Hennig et al. 1975). Thus, the comparison with Hess's cytogenetic map (Hess

1965b, Fig. 18) reveals only minor differences. Th and Ps are closer together and located more proximally. Most probably, the Tr and the Cl are also located close to each other in a more distal location than given by Hess. As in *D. melanogaster*, large blocks of the Y chromosome are devoid of male-fertility genes. This does not mean, however, that these regions do not have a function. At least for *D. melanogaster* several functions of such a region have been demonstrated (see below).

3.6 Lampbrush Loops Formed by the Y Chromosome of *D. hydei* and *D. melanogaster*

Characteristic, species-specific, Y-chromosome lampbrush loops are found in the primary spermatocytes of all of the 54 *Drosophilids* which have been investigated thus far (Hess 1967b; Hess and Meyer 1968). In many species, such as *D. melanogaster* and *D. simulans*, the lampbrush loops are inconspicuous in phase-contrast microscopy, whereas in other species they dominate the nuclear morphology. In *D. hydei* and some related species the lampbrush loops are particularly prominent and have a characteristic morphology (Meyer 1963; Hess and Meyer 1963a, b). Five pairs of lampbrush loops can be distinguished in *D. hydei* (cf. Sect. 3.4 and Fig. 12). Cytological evidence shows that all four loops of the long arm are connected directly and form an open chain (Fig. 12b and c). There are no indications that parts of the Y chromosome are included into the nucleolus as suggested by Hess and Meyer (1968) and Hess (1980). There are also no indications for a chromomeric structure of the Y chromosome, where lateral extrusions from condensed parts of the Y chromosome are formed which give rise to the lampbrush loops (Kremer et al. 1986).

The DNA axis within the lampbrush loops has recently been visualized with the aid of the DNA-specific dye DAPI (Kremer et al. 1986). The loop architecture is quite complex; differential staining reactions show that the loops are composed mainly of protein and RNA (Yamasaki 1977, 1981). At the elecron microscope level the ultrastructure of the different loops is characteristic for each loop (Grond 1984; Grond et al. 1984). These results are contrary to the statement of Hess (1980) that there is no discrimination possible between the different loops at the ultrastructural level. Histochemical reactions, such as uranyl staining at different pHs, PTA staining, and EDTA bleaching have shown that, in particular, the loops Cl, Ps, and Th contain huge amounts of proteins, but are poor in nucleic acids (Grond et al. 1984). Monoclonal antibodies which recognize RNP react with several lampbrush loops (Glätzer 1984; Glätzer and Kloetzel 1985). But the existence of loop-specific proteins was questioned by Hess (1980) and Kloetzel et al. (1981). However, the existence of an 80000 dalton protein in the pseudonucleolus (Ps) has been demonstrated with the aid of an antiserum, which specifically decorates this lampbrush loop (Hulsebos et al. 1983, 1984). This loop also contains a protein which reacts with antibodies against histone H1 (Kremer et al. 1986). The reaction of the serum with testicular proteins shows that the 80000 dalton protein is coded by the X chromosome or the autosomes. A protein of $Mr = 35000$ has been demonstrated by a monoclonal antibody to be specific to the loop Ns (Hennig et al., unpublished).

Fig. 15. The kl-3 loop (*arrowheads*) of *D. melanogaster* in indirect immunofluorescence. Primary spermatocytes of wild-type X/Y males were prepared by testis squashes and incubated with the anti-hsp155 serum (Hulsebos et al. 1984). Treatment with a second fluorescent antibody decorates specifically the kl-3 lampbrush loop. *Bar* represents 10 µm. (Photograph by W. Hennig)

The 80 000 dalton protein of the Ps loop of *D. hydei* is rather conserved. It decorates nuclear structures in the primary spermatocytes in almost all other *Drosophila* species including *D. melanogaster* (Hulsebos et al. 1984). In *D. melanogaster* a giant lampbrush loop becomes visible (Fig. 15 and Fig. 4a of Hulsebos et al. 1984). With the aid of our antiserum, Bonaccorsi and Gatti (personal communication) were able to demonstrate that this loop is formed by kl-3. As in *D. hydei* (Hackstein et al. 1982) male-sterile mutations in the loop-forming gene remove the loop, modify the loop, or do not affect the loop. Therefore, it is evident that also in *D. melanogaster* male-fertility genes form giant lampbrush loops in the primary spermatocytes. This could already be shown by Kiefer (1973) in

Fig. 16 A–C. Lampbrush loop nooses (*Ns*) of *D. hydei*, visualized with the Miller spreading technique (Grond et al. 1983). **A** The length of the loop is 47.5 µm (*thick arrows*); the DNA content is estimated to be 260 kb. There is only one initiation point within the loop (*right side*), and the transcripts increase in size over the entire length of the loop. *Insets 1* and *2* display some sections of the loop with higher magnification (*arrows*). In *inset 1* the DNA axis is visible, and in *inset 2* a dislocated ribosomal RNA gene is shown. **B** A primary spermatocyte nucleus in phase-contrast from a T(X;Y)58/O male. Only the nooses (*Ns*) are seen because of the deletion of the long arm of the Y chromsome. Such nuclei were subjected to Miller spreading in order to obtain a picture as displayed in **A**. **C** A Giemsa-stained metaphase plate from a larval neuroblast from a T(X;Y)58/O male. The X chromosome carries a translocation of the short arm of the Y chromosome (*arrow*). *Bar* represents 10 µm. (Hennig 1985a and Grond et al. 1983)

Spermatogenesis in *Drosophila*

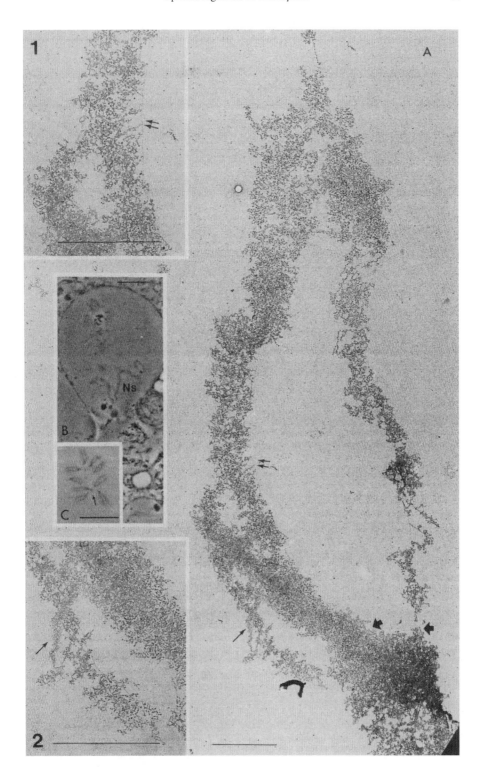

phase-contrast photographs. Also, for the fertility factors kl-5 and ks-1 of *D. melanogaster* the formation of lampbrush loops could be demonstrated with the aid of indirect immunofluorescence (Melzer and Glätzer 1985).

For *D. hydei* it had been recognized early that the lampbrush loops must contain a rather long stretch of DNA (Hess and Meyer 1968); measurements of Hennig et al. (1974a) showed that the loop length of the different lampbrush loops varied from a minimum of 10 µm to a maximum of 60 µm. This implies that each loop contains enough DNA to code for more than 100 polypeptides (Hennig 1978). Genetic evidence in *D. hydei* shows that each lampbrush loop contains only one complementation group. This could mean that each lampbrush loop gives rise to only one transcript. These transcripts must be longer than 100 kb. The only way to demonstrate such large transcripts is by visualization using the Miller spreading technique (Miller and Beatty 1969). Hennig et al. (1974a) and Meyer and Hennig (1974a, b) were the first to spread chromatin from primary spermatocytes and to demonstrate the existence of giant transcripts. Glätzer (1975) and Glätzer and Meyer (1981) obtained perfectly spread transcripts from X/Y testes. They demonstrated that in the testicular cells, most likely primary spermatocytes, a number of different giant transcripts can be found. They came, however, to the conclusion that there are no loop-specific transcripts. Grond et al. (1983) used a translocation stock which carries only the lampbrush loop nooses [T(X;Y)58/0 males]; the other lampbrush loops are deleted in this stock. They succeeded in the spreading of the complete lampbrush loop nooses (Fig. 16). More than 47.5 µm

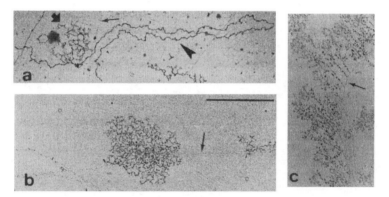

Fig. 17 a–c. Transcripts from the lampbrush loops threads (**a**), pseudonucleolus (**b**), and nooses (**c**) (Grond et al. 1983; deLoos et al. 1984). Primary spermatocyte nuclei from translocation-bearing males were subjected to the Miller spreading technique and prepared for electron microscopy. Each loop displays a highly characteristic secondary structure in the transcripts and in the distances between the transcripts. The transcripts of the threads (**a**) are composed of a bushlike part (*thick arrow*) and a long, distal part (*arrowhead*). The RNA in this threadlike part is thicker than the RNA in the bushlike region, and therefore, probably associated with protein. The pseudonucleolus (**b**) contains only bushlike transcripts which are much larger than the bushlike part of the thread transcripts. Often smaller transcripts (to the *right* of the *arrow*) are found between the great bushlike transcripts, indicating secondary initiation sites. Such secondary initiation points have not been found in the other loops. The distance between the individual transcripts in the threads and in the pseudonucleolus is large if compared with the nooses (**c**). The noose transcripts initiate more frequently than transcripts of the other loops. The *arrows* in **a, b, c** indicate the DNA axis. *Bar* represents 1 µm. (Hennig 1985a)

DNA are covered by growing transcripts. Approximately 140 transcripts are found attached to the DNA axis with an average intertranscript distance of 0.35 µm. If one assumes a DNA compaction ratio of 1.6, then more than 260 kb DNA are transcribed in a single transcription unit. The transcripts have a highly complicated secondary structure, which may reflect the repetitive sequence organization of the loop (see below). The branched RNP fibers of the transcripts are covered with granules of a diameter of 39 ± 10 nm. Since the granules are evenly spaced, each transcription unit can carry 7500 to 16000 granules.

Similar results have been obtained with translocations of Y fragments that carry only the threads [T(X;Y)6/0], the pseudonucleolus [T(X;Y)74/0], or the clubs (X/YK-173-3). In all three cases, loop-spedific transcripts have been found (de Loos et al. 1984; Fig. 17). The size of the transcript for the threads is between 500 and 1000 kb; for the pseudonucleolus, about 1500 kb (de Loos et al. 1984), and for the clubs, between 500 and 1000 kb (Hennig and Suijkerbuijk, unpublished). The transcripts have a highly complicated secondary structure. Therefore, four of the five loops (male-fertility genes) produce giant transcripts of a size between 260 and 1500 kb with a loop-specific secondary structure. The same situation may also pertain for *D. melanogaster*, where giant lampbrush loops are also formed. Cytogenetic studies on metaphase Y chromosomes also indicate a large size of the male-fertility genes in this species (Sect. 3.3), since the breakpoints that inactivate a fertility gene can be scattered over a wide range of the Y chromosome.

3.7 Y Chromosome Mutations and Spermatogenesis

For many years the function of the Y chromosome has been investigated by the analysis of the morphological effects caused by the lack of the Y chromosome. Bridges (1916), Safir (1920) and Shen (1932) found that X/O males of *Drosophila melanogaster* were sterile and were devoid of motile sperm. Stern (1929) showed that the deletion of only one of the two chromosome arms of the Y chromosome was sufficient to render males sterile. In 1960 Brosseau demonstrated that the fertility complexes of *D. melanogaster* can be subdivided into seven fertility genes: the mutation of even a single male-fertility gene sterilizes the males as effectively as the loss of the complete Y chromosome. The sterility is accompanied by an interruption of the process of sperm maturation (Brosseau 1960). Cytological studies of spermatogenesis in Y-deficient males have been undertaken by a number of authors. Bairati and Baccetti (1966); Kiefer (1966, 1968, 1969, 1970, 1973); Meyer (1968, 1972) and Hess and Meyer (1968) tried to determine the morphological basis for the sterility of those males which lack the complete Y chromosome or a single fertility gene. The objective of all of these studies was to correlate specific lesions of sperm morphogenesis with Y-chromosome function. The defects which were found can be grouped into four categories: (1) irregularities in the pattern or the number of axoneme microtubules, (2) abnormalities in the development of the nebenkern derivatives, (3) failures of the individualization of the sperm, and (4) reduced numbers of spermatids per cyst.

Axonemal defects are found frequently, but there is great variability between different cysts. In addition, there is controversy over the significance of axonemal defects. Meyer (1968) showed axonemal defects which he described as typical defects for X/O spermatogenesis. The same defects, however, were interpreted by Kiefer (1973) as characteristic for degeneration of spermatids for any of a variety of causes. Therefore, it is sometimes difficult, if not impossible, to distinguish between effects caused by developmental failure and those which are due to degeneration. Degeneration of sperm is a prominent phenomenon in X/O and X/ms(Y) males because the spermatids do not mature, but degenerate. Therefore, at the same time, the testes of sterile males contain developing and degenerating cysts.

Typical for nebenkern defects is the failure of the proper association of the nebenkern derivatives with the axoneme. The formation of the paracrystalline material, which in *D. melanogaster* normally appears only in the larger of the two nebenkern derivatives, can also be severely disturbed. But it is important to stress that all necessary structural elements are apparently present in X/O spermatids. The structural defects, therefore, can be interpreted as organizational failures.

The third type of defect, failure of individualization, is found frequently when spermatid development proceeds rather far. Severely disturbed stages of individualization of spermatids can be found, in which the spermatids remain in their syncytial state and do not spiralize.

The cysts of X/O males nearly always contain less spermatids than normal; the mean value is 31 spermatids per bundle in contrast to the 64 present in X/Y males (Kiefer 1966). It has been shown that X/O males behave as meiotic mutants which fail in the regular segregation of the chromosomes (Lifschytz and Hareven 1977; Lifschytz and Meyer 1977). Male meiosis is controlled by an Y-chromosome site near kl-2 (see below).

Kiefer (1973) tried to correlate the appearance of the different effects with the absence of individual Y-chromosome loci. All defects can be found in nearly all the different genetic constitutions, with however slightly different frequencies. A ranking of the Y mutants relative to the overall disturbance of spermiogenesis was possible. The kl-1$^-$ constitution had a minimum of effects, ks-1$^-$ and ks-2$^-$ most closely resembled the X/O phenotype.

For *D. hydei* a discrimination between "early" and "late" effects has been made (Meyer 1968; Hess and Meyer 1968). The different effects have been correlated with the presence or absence of certain lampbrush loops. There is, however, a lack of a well-defined staging of spermatogenesis; therefore, the discrimination between early and late effects remains arbitrary. The lack of the threads or the pseudonucleolus (or both) leads to late effects with nearly normal sperm; a deletion of the whole long arm or of the short arm of the Y chromosome causes early effects which resemble an X/O phenotype. The correlation of a lack of Th, Ps, and Tr with early effects by Hess and Meyer (1968) cannot be confirmed by us; the lack of the short arm or a male-sterile mutation in locus Q, however, invariably leads to an X/O phenotype (Hackstein et al. 1982). X/O males of *D. hydei* are similar to *D. melanogaster*, perform meiosis, and show a reasonable differentiation of the spermatids (Hennig et al. 1974b). It has been claimed by Hess and Meyer (Hess and Meyer 1968; Meyer 1968) that in contrast to *D. melanogaster*, X/O males of *D. hydei* are blocked before meiosis and completely lack postmeiotic stages. Another statement of Hess and Meyer can also not be confirmed. They described a correlation between the length of the sperm and the amount of Y chromosome material in *D. hydei*. X/Y/Y males were reported to contain sperm of a length twice (± 12 mm) that of a normal X/Y male (± 6 mm). For *D. melanogaster* such a correlation could not be confirmed, though this species has a much smaller sperm size (± 1.8 mm) (Lindsley and Tokuyasu 1980). In normal X/Y *D. hydei* males a sperm size of more than 12 mm has been measured (Grond and Hennig, unpublished). No recent information about the sperm size of X/Y/Y males of *D. hydei* is available.

The conclusion of all these morphological analyses of Y deletions in *D. melanogaster* has been clearly formulated by Kiefer (1973): "Our studies on the effects of various Y mutations on spermiogenesis have been disappointing in light of our original objectives. It has not been possible to relate the presence of a specific structural or developmental lesion to the absence of a specific Y locus."

In 1981 Hardy et al. undertook a new attempt to correlate the lack of individual male-fertility genes with the earliest abnormality in the spermatogenetic process, consistently associated with the lack of that factor. Deletions of specific male-fertility factors were produced by the method of segmental aneuploidy, which had been introduced by Kennison (1981) (Figs. 3 and 4). Their stocks allowed the generation of seven adjacent, nonoverlapping deletions: six regions carry a male-fertility gene, and one (E) is devoid of a fertility gene (see Fig. 4 for explanation of A, B, C, D, E, F, and G). Spermatogenesis of males deficient for one such region was examined with the light and the electron microscope from the primary spermatocyte stage through the formation of mature sperm. The lack of region A (containing kl-5) or region B (containing kl-3) caused the first abnormalities in the primary spermatocyte stage. The tubular or reticular material, respectively, in the nuclei was largely reduced. This material has been identified by Meyer et al. (1961) as belonging to the „Funktionsstrukturen" of the Y chromosome. Therefore, it is likely that the lack of tubular or reticular material simply reflects the absence of the kl-5 or kl-3 lampbrush loop. Since in the deficient males some tubular or reticular material is still found, this material should be synthesized outside the two loci, most likely outside the Y chromosome. For the kl-3 loop it has been shown that the loop-specific protein was synthesized in *D. hydei* also in an X/O constitution but did not bind to a loop (Hulsebos et al. 1984). Another abnormality which is induced by the absence of each of the regions A and B (kl-5 and kl-3) is the lack (or reduction) of one of the projections which extend from the A subtubule of each peripheral doublet of microtubules in the axoneme. This projection or "outer arm" is suspected to contain dynein, a high molecular weight protein constituent of cilia and flagella which has an ATPase activity. Therefore, it is possible that the deletion of kl-5 and kl-3 both interfere with the formation of the dynein arm of the sperm flagellum. Apparently both deletions cause the same phenotype. In later stages of development the spermatids with deletions of kl-3 and kl-5 encounter difficulties in the individualization which have been described by earlier investigators. It remains unclear whether these different effects belong to the same cascade of events which are caused by the deletion of male-fertility factors or whether they are unrelated.

The deletion of region C (containing kl-2) leads to a phenotype very similar to a typical X/O phenotype. Crystals are formed in the nuclei and the cytoplasm of the primary spermatocytes. Meiosis is grossly abnormal, and tripolar and tetrapolar spindles can be observed as in X/O individuals (cf. Lifschytz and Meyer 1977). Multiple nuclei and nebenkerns of different size are formed. It has, however, recently been shown that all these effects are not caused by the lack of kl-2, but by the deletion of an adjacent male-fertile locus (see below). Therefore, the function of kl-2 and the phenotype of kl-2 deficient males remain unclear.

In males deficient for region D (kl-1) sperm development was hardly discriminable from normal males. In the individualization stage, however, a cross-section through a cyst revealed a great variation in the shape and the cross-sectional area of the nebenkern derivatives. Some spermatids showed an abnormal, hourglass-shaped, cross-sectional profile. During the coiling process the heads of the more abnormal spermatids slipped out of the head trap and became displaced caudally. Therefore, the spermatids of a bundle got out of register, which accounted for the variability in the cross-sectional area of the mitochondrial derivatives. Also, the minute tubules which pack the cyst lumen at the time of coiling were absent. There existed, however, a kl-1 mutant, which exhibited regular coiling and which contained these minute tubules. Most likely this mutant was a hypomorph, which differed in some aspects from the null mutant which was formed by the deletion of region D.

The deletion of region F(ks-1) leads to a phenotype which was very similar to that of kl-1$^-$. More frequently, however, cross-sectional profiles were found which exhibited an hourglass shape. The coiling appeared to be nearly normal with only a few sperm out of register.

The first abnormalities in the development seen in males deficient for region G (ks-2), appeared in the onion nebenkern stage. The axoneme, which was ensheathed at this stage, was found in an abnormal position. It was displaced laterally from the furrow which was formed at the junction of both nebenkern halves.

In conclusion, deletions of kl-5 and kl-3 both prevent the formation of the outer dynein arm of the axoneme. The deletion of ks-2 disturbs the proper apposition of axoneme and nebenkern. The deletion of kl-1 or ks-1 leads to late effects and the deletion of kl-2 does not allow the identification of an early phenotype. Therefore, only for a part of the Y-chromosome loci an early deviation from the normal development can be found. Hardy et al. hoped to find single, consistent, and unifying correlations between the lack of certain male-fertility genes and the early effects on spermatid differentiation. Their aim was to demonstrate that the Y-chromosome factors behave in an "orthodox manner characteristic of structural genes found on the other chromosomes." Unfortunately, this objective in my opinion is not completely fulfilled. The lack of the outer dynein arm is a well-definied phenotype, but it is caused by deletion of kl-5 as well as by deletion of kl-3. Deletion of kl-1 causes a rather diffuse phenotype which is shared by deletion of ks-1. Thus, both phenotypes are caused by more than one mutation. The phenotype of ks-2 seems to be the only clear deviation from the normal situation which can be correlated with a single Y-chromosome site; but it remains unclear whether the proper apposition of axoneme and nebenkern can be controlled by a single protein. Therefore, the role of the Y chromosome in spermatogenesis remains unclear and open for new interpretations.

It has already been mentioned that the deletion of Y-chromosome material around kl-2 leads to a complex phenotype with nuclear crystal formation and meiotic abnormalities. The heterochromatic regions h 10, h 11, h 12, and h 13 (Figs. 10 and 11) could be removed between the breakpoints of the male-fertile translocations W 27 and V 31 (Hardy et al. 1984). The analysis of the effects of deletions generated by the combination of nine additional male-fertile translocations with breakpoints between W 27 and V 31 revealed that kl-2 could be located in the nonfluorescent block h 10. Deletion of h 10 renders the males sterile and removes one high molecular weight protein (see below). There is, however, no crystal formation and normal meiosis takes place. The deletion of regions h 11, h 12, and h 13 does not sterilize normal males, but leads to crystal formation and an abnormal meiosis. The segregation of all chromosomes except chromosome 4 is affected and the meiotic chromosomes have a sticky appearance. It had been speculated that this Y-chromosome region contains a collochore (Cooper 1964) which controls the segregation of the X and Y chromosome. Since the segregation of autosomes 2 and 3 is also affected, the segregational distortion must have other causes.

The shape of the nuclear crystals (needle-shaped vs star-shaped) is controlled by an X-chromosome site, Ste (Stellate, 1-45,7). Ste is dominant over the Ste$^+$ allele which is present in normal stocks (Meyer et al. 1961; Hardy 1980; Hardy et al. 1984). A deletion of the Y-chromosome blocks h 11-13 in the presence of an X-chromosome Ste$^+$ leads to the formation of needle-shaped crystals in the spermatocyte nuclei and cytoplasm. In the presence of Ste, star-shaped crystals are formed. Therefore, Y-chromosome sites control the expression of an X-chromosomal gene. These results can be correlated with biochemical data. An mRNA, which is abundant in testes of X/O males, but not abundant in testes of X/Y males has been isolated by Lovett (Lovett et al. 1980; Lovett 1983). This RNA sequence hybridizes to the X chromosome at 12F1-2, which is the region containing the Ste locus. The synthesis of this messenger is suppressed by a Y-chromosome region between kl-1 and kl-2. The messenger is translated into a 17 000 dalton polypeptide in a cell-free system. It has been shown that one clone (Dm 2L1) which is homologous with this messenger contained about eight copies of a 1250-bp sequence, called 2L1 (Livak 1984). Ste-bearing X chromosomes contain about 200 copies of this sequence, whereas Ste$^+$ chromosomes contain only about 20 copies. Sequences homologous to these X-chromosome 2L1 sequences are present also on the Y chromosome in region h 11 ($\pm$80 copies). Thus, there are sequences which are shared by the X and the Y chromosome. Ste X chromosomes contain a high copy number, Ste$^+$ a low copy number. The sequences on the Y chromosome most likely suppress the transcription of the X-chromosome copies; a deletion of the Y-chromosome copies leads to excessive transcription of the X-chromosome sequences and subsequently to the accumulation of protein in the crystals with the Ste-dependent morphology.

It is still unknown if the Y chromosome codes for structural proteins of the sperm (Hennig et al. 1974a). Two-dimensional gel electrophoresis failed to reveal any difference in the polypeptide pattern between X/Y and X/O males in *D. mela-*

nogaster (Ingman-Baker and Candido 1980; Kemphues and Kaufman 1980). In contrast to these observations, Goldstein et al. (1982) demonstrated that the deletion of kl-2, kl-3, and kl-5 results in the absence of three high molecular weight proteins from testes of *D. melanogaster*. These proteins have molecular weights in the 300 000–350 000 dalton range; their mobilities are similar to those of the dynein polypeptides from *Chlamydomonas* axonemes. ^{35}S Methionine-labeling revealed a positive relationship between the doses of a given fertility gene and the concentration of the corresponding protein in the testis. However, this relationship is not linear, and furthermore, the deletion of that gene does not result in the complete absence of that protein. A number of other proteins also shows variation in concentration. Nevertheless, Goldstein et al. (1982) interpreted their results in favor of a protein-coding function of the Y-chromosome loci. For *D. hydei* it has been shown that the concentration of different testis-specific proteins is under the control of Y-chromosomal genes (Hulsebos et al. 1982, 1983). The deletion or the inactivity of the Y-chromosomal genes O, P, Q leads to the disappearance of the proteins sph 155 and sph 35; the concentration of tubulins is also greatly reduced. All these proteins are most likely encoded by genes outside the Y chromosome; their concentration is regulated by the Y-chromosome loci. It has been shown for the tubulin messenger RNAs that their concentration is not changed; if one compares X/Y and X/O individuals, a normal level of tubulin messengers is present in X/O testes (Brand et al., unpublished). Therefore, with the current state of knowledge it is impossible to discriminate between Y-encoded and Y-regulated proteins by the analysis of protein gels. It remains an open question whether the Y chromosome codes for structural proteins of the sperm or not.

3.8 DNA Sequences from the Y Chromosome of *D. hydei*

For many years numerous attempts have been undertaken to isolate Y-chromosome DNA sequences (Hennig 1972a, b; Renkawitz 1978a, b, 1979). These attempts were not very successful until recombinant DNA techniques allowed new approaches. Four different strategies have been used to isolate Y-chromosome sequences. The first method allows the recovery of Y-specific sequences or sequences which are predominantly Y-chromosomal by a quantitative dot blot hybridization screen: a library of genomic DNA is hybridized successively with labeled DNA from males and females (Lifschytz 1979; Vogt et al. 1982; Vogt and Hennig 1983; Lifschytz et al. 1983). If hybridization of a sample is achieved only with DNA from males (X/Y) and not with DNA from females (X/X), then the sequence is presumed to be localized on the Y chromosome. On the other hand, if a sample reacts with both "male" and "female" probes, then this sequence is localized on the other chromosomes, but not excuded from the Y chromosome. A quantitative difference between the hybridization with DNA from males and with DNA from females indicates that a sequence is located on the Y chromosome, but also on the X chromosome or the autosomes. The second method for the isolation of Y-chromosome sequences depends on the isolation of testis-specific RNAs and the construction of a cDNA library. This library is screened differentially with labeled testes and carcas RNA, and subsequently testes RNA-specific clones are used as a probe for Southern blot hybridization to DNA from males and females in order to identify transcribed, testis-specific, Y-chromosome DNA sequences (Brand and Hennig 1987). The third strategy depends on the manual isolation of Y-chromosome lampbrush loops from primary spermatocytes and their microcloning in lambda-vectors (Hennig et al. 1983). The fourth method, a technical variant of method 1, uses DNA from females bound to a macroporous support in chromatographic columns in order to enrich Y-chromosome sequences (Bünemann 1982; Bünemann et al. 1982; Awgulewitsch et al. 1985).

The analysis of the Y-chromosome DNA sequences which were obtained by these four methods revealed that they belong to two different classes (Hennig 1985b). The one class of DNA sequences is Y-specific, i.e., it is restricted in its genomic location to the Y chromosome. The other class of sequences is Y-associated; these sequences are located on the Y chromosome, but they occur in other locations (X chromosome, autosomes) as well (Vogt and Hennig 1983; Hennig et al. 1983; Huijser 1985; Hareven et al. 1986). The location of these sequences can be unequivocally determined with the aid of hybridizations in situ to polytene chromosomes, metaphase chromosomes, and primary spermatocytes.

The Y-specific DNA sequence class is organized in repeat families of 200-2000 copies. They are typically arranged in tandem repeats; the length of one copy does not exceed a few hundred bp (Vogt 1985; Vogt and Hennig 1986a). One repeat family is restricted in its location to single (Lifschytz et al. 1983; Awgulewitsch et al. 1985) or functionally related lampbrush loops (Huijser and Hennig 1987). This can be demonstrated by transcript-in situ hybridizations to the lampbrush loops of primary spermatocytes: a labeled cDNA probe from Y-chromosome clones complementary to a loop-forming site hybridizes with the growing RNA transcripts which cover the DNA axis of the lampbrush loop. DNA sequence analysis of one repeat family (from the lampbrush loop nooses) reveals no indication for a protein-coding capacity (Vogt and Hennig 1986a). It is, however, evident that the sequence has a high capacity to form secondary structures due to internal, direct, and inverted repeats. A part of this sequence shares homologies with the ARS ("autonomously replicating sequences") sequences of yeast; another part is homologous to those DNA sequences which promote the amplification of the chorion genes of *D. melanogaster*. Also, sequence homology to enhancers of transcription are found. It is assumed that the accumulation of these sequences within the lampbrush loop is not accidental, but is connected to the function and the evolution of the lampbrush loop. During the evolution of a lampbrush loop, amplification events must be very important.

The Y-associated sequences are characterized by a smaller family size; usually not more than 50 copies per genome are found. The lengths of the DNA sequences are highly variable and may reach several kb. Only some copies of such a gene family are Y-chromosomal. The other copies reside on the X chromosome and the autosomes (Vogt and Hennig 1983, 1986b). Often only in situ hybridizations demonstrate the Y-chromosome location unequivocally. Their autosome and X-chromosome locations vary considerably in closely related species (*D. neohydei, D. eohydei*) or even in wild-type strains of different origins (*D. melanogaster*). Therefore, it is assumed that these Y-associated sequences are transposable elements which became incorporated into the Y chromosome recently; in some species these sequences are exclusively autosomal or X-chromosomal (Hareven et al. 1986; Vogt et al. 1986). On the Y chromosome these sequences are interspersed with the Y-specific sequences (Vogt and Hennig 1986 b). Both sequences are transcribed as a single, primary transcript from a giant transcription unit. From the Y-associated sequences discrete size classes of RNA can be recovered (Brand and Hennig 1987), whereas from the Y-specific class only RNAs of heterogeneous size are found (Vogt et al. 1982).

The evolutionary conservation of both classes of Y-chromosome sequences is poor (Hareven et al. 1986; Vogt et al. 1986). The occurrence of the Y-specific class is restricted to the sibling species *D. neohydei* and *D. eohydei,* with a remarkable loss of homology even in *D. eohydei.* The distribution of the Y-associated class is broader; these sequences share homologies with more distant *Drosophilids,* but frequently only the non-Y-chromosome locations are conserved. Evidently there is a rapid evolution of the Y-chromosome DNA sequences. This process starts most probably with the fixation of a new Y-chromosome sequence; subsequently, this sequence becomes amplified and diverges rapidly in order to give rise to a large block of closely related, tandemly repeated sequences. This fact finds its cytological expression in the highly variable, but species-specific cytology of the Y-chromosome lampbrush loops in the primary spermatocytes of many *Drosophilids* (Hess 1967 b).

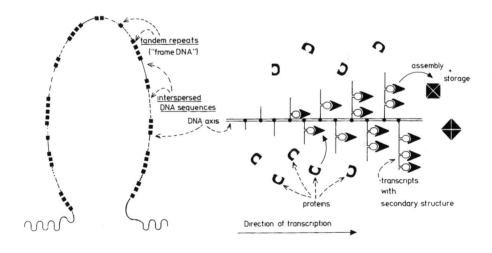

Fig. 18. A model of the structure and function of a typical lampbrush loop of *D. hydei.* The lampbrush-loop DNA is built up from a frame of tandemly repeated Y-specific sequences. Interspersed between these Y-specific sequences are several different, moderately repeated, Y-associated DNA sequences. Transcription of the loop DNA in a single transcription unit generates a gradient of growing RNA molecules. The loop-specific secondary structure facilitates the binding of loop-specific proteins by the transcripts. The bond proteins assemble and are stored within the lampbrush loop matrix. (Courtesy W. Hennig)

The sequence organization and the lack of open reading frames in the DNA sequences of the Y-specific class make it very unlikely that the lampbrush loops have a protein-coding function. The lampbrush loops accumulate specific proteins (see Sect. 3.6) and their DNA sequences are suited for protein binding due to their high degree of secondary structures (Hennig et al. 1987). These proteins occur not only attached to the primary transcripts, but are also accumulated in the superstructure of the lampbrush loop (Fig. 18). It is proposed that these proteins play an important role in the postmeiotic nuclear differentiation of the spermatid which is characterized by several chromatin condensation-decondensation cycles (cf. Kremer et al. 1986). Thus, the male-fertility genes have an exceptional size, structure, and function. Whether they still have some protein-coding capacity next to their protein-binding properties remains to be proven.

4 Conclusions

Spermatogenesis has three main genetic requirements: first, it needs the coordinate expression of genes which code for the structural components of the sperm; second, it needs a set of genes which control the expression of the structural genes; third, it depends on a certain structural organization of the genome, i.e., it is sensitive to chromosome rearrangements. The structural genes must be sought under the X-chromosomal and autosomal genes which can mutate to male sterility. The number of male-sterile mutations has been estimated to lie between 600 and 2400 (Lindsley 1982). The lower number is an extrapolation based on the relative frequency of male-sterile mutations, the higher estimate accounts for the extrapolated number of ts lethals which are sterile if raised under permissive conditions. A third of all ts lethals are male-sterile under conditions which allow the survival of the imagos. This is a strong indication that nearly every metabolic stress can cause male sterility. Even minor deviations from normal metabolism, as it is the case in ts lethals under permissive conditions, are sufficient to render males sterile. Therefore, I believe that the vast majority of all male-sterile mutations cause male sterility merely by the pleiotropic action of genes unrelated to spermatogenesis. Thus, it is no trivial task to identify genes which code for structural components of sperm. The great challenge for the future is the design of screening procedures which allow the recovery of mutations specific for spermatogenesis. This is surely not accomplished by screening simply for male-sterile/female-fertile mutations, since it is very likely that spermatogenesis is much more sensitive to metabolic stress than oogenesis as the high number of male-sterile ts lethals indicates. A screening procedure has to depend on criteria which allow the discrimination of specific morphological deviations from normal spermatogenesis. Alternatively, one has to screen for structural genes which code for biochemically defined sperm components. There are only two cases in which a spermatogenesis-specific protein and the corresponding gene have been identified. The first example is the Ste gene on the X chromosome of *D. melanogaster* which is suspected to code for a protein that forms the crystals in the primary spermatocytes of X/O males. In X/Y males this protein seems to have a function in meiosis (Hardy et al. 1984). However, the exact gene-protein-function correlation remains to be established. The second example is the B2t locus on the third chromosome of *D. melanogaster;* this locus is the structural gene for the testis-specific β_2-tubulin (Kemphues et al. 1979, 1983). All mutations in this site sterilize males; certain alleles exhibit a complex phenotype. The meiotic spindle, the spermatid nucleus, and the sperm flagellum are the targets of mutations in the B2t locus, since microtubules are present in the meiotic spindle, microtubules shape the spermatid nucleus, and microtubules are constituents of the sperm flagellum. The B2t[8] allele, which leads to morphologically aberrant microtubules, unequivocally demonstrates that this mutation in the structural gene causes all three defects (Fuller 1986), indicating that β_2-tubulin participates in all three structural arrays of microtubules. The product of the B2t interacts with the products of other genes, which most likely code for functionally related components of the microtubule complexes. These genes, so-called second-site noncomplementors, are sterile if combined in trans with certain B2t alleles (Fuller 1986). These peculiar properties make it possible to screen systematically for genes which interact with the testis-specific β_2-tubulin.

Almost certainly the collective of autosome and X-chromosome male-sterile sites also contains regulatory genes. However, it will be difficult to identify these mutations. The Y-chromosomal male-fertility factors are most likely regulatory genes. Mutations in the majority of the Y-chromosomal male-fertility genes do not cause a specific phenotype and no deletion of a fertility factor leads to the absence of major structural sperm components (cf. Hardy et al. 1981 and Sect. 3.7). Though it has been claimed that Y-chromosomal genes code for structural proteins (Goldstein et al. 1982), the evidence for this is poor, and the results presented in this chapter also permit an interpretation in favor of a regulatory role of the Y chromosome. DNA sequence analysis failed to demonstrate open reading frames and, therefore, also argues against a protein-coding function. This does not exclude protein coding from widely separated small exons. Indeed, the regulatory function of the Y-chromosome sites may be exerted on the protein, and not on the RNA level. It is possible that proteins which are engaged in the regulation are present only in catalytical quantities and, therefore, are missed in conventional two-dimensional protein gels. These proteins can be minor structural components which direct a self-assembly of the major constituents of the sperm or they can be regulatory molecules which control the activity of the structural genes.

Another function of the lampbrush loop-forming, Y-chromosomal, fertility genes is very likely: protein binding. It has been demonstrated in *D. hydei* that the lampbrush loops accumulate large amounts of proteins which may be important for the postmeiotic development of the spermatid, perhaps in a possible histone transition. However, the protein binding by Y-chromosomal transcripts has to be proven unequivocally by suitable experiments, and the postmeiotic fate of the proteins has to be chased very carefully. However, it is very probable that protein binding is not the only function of the loop-forming fertility genes, since mutations can very frequently abolish the gene function. The high frequency of Y-chromosome, male-sterile mutations makes it probable that nearly every mutational event along the gene sterilizes the male. The protein-binding function should be rather tolerant against single base substitutions, and it is improbable that the binding capacity is such a critical parameter that stop mutations close to the end of the transcription unit or that small internal deletions interfere with the function. Thus, the high mutation frequency argues that the loop-forming genes have additional functions besides protein binding and that the whole length of the genes is sensitive to mutation damage.

The high mutation rate and the small number of Y-chromosomal genes create a paradox which can be solved if one takes into account the large sizes of the fertility factors. The extreme size of Y-chromosomal transcripts in *D. hydei* has already been demonstrated by Hennig et al. (1974a) and Meyer and Hennig (1974b). Their locus-specific secondary structure and their dimensions have been elucidated by Grond et al. (1983) and de Loos et al. (1984). Cytogenetic analysis of mitotic Y chromosomes of *D. melanogaster* has shown that the function of the male-fertility genes depends on the integrity of large segments of the Y chromosome (Gatti and Pimpinelli 1983). A large size of the genes is one possible interpretation; however, with the current state of knowledge, distance – or position – effects acting over a long range cannot be excluded.

The possible reasons for the enormous dimensions of the male-fertility genes remain completely unclear. It also remains an enigma whether genes with similar dimensions and properties are present on the X chromosome or the autosomes, since the existence of such genes cannot be demonstrated easily.

The cytogenetic analysis of mitotic Y chromosomes of both species *D. melanogaster* and *D. hydei* demonstrated that large sections of the Y chromosome are devoid of male-sterile genes. Only for some of the regions it has been shown that they are dispensable for normal spermatogenesis. For the other regions it still remains to be demonstrated whether they have functions specific for the germ line or whether they represent inert genetic material which is dispensable in both the germ line and the soma (cf. Hennig 1986).

From a superficial point of view, the genetics of the Y chromosome of *D. melanogaster*, and in part also of *D. hydei*, might be considered as a solved problem, even with regard to the number of sites. It is true that the method of segmental aneuploidy has proven the limited number of fertility genes on the Y chromosome of *D. melanogaster*. However, the number of six sites in *D. melanogaster* is lower than the number of 16 fertility genes on the Y chromosome of *D. hydei*. The number of sites in *D. hydei* may be an overestimation (see Sect. 3.4), but on the other hand, there is no necessity for an equal number of fertility genes in both species. Besides, the number of sites in *D. melanogaster* might be higher. Since the frequency of induced mutations in the known Y-chromosomal fertility genes is very high, only moderate numbers of Y chromosomes have been screened. Therefore, it is reasonable to assume that the Y chromosome is not exhaustively screened and saturated for male-sterile mutations. Especially genes with a low mutation rate and a location close to the known genes might have escaped their detection. Unsolved is also the problem which is generated by the unstable, male-sterile mutations recovered in $sc^{4L}sc^{8R}/Y^*$ males (see Sect. 3.2) and the leakiness of a number of Y-chromosome sites in *D. melanogaster*.

It is inherent to the screens (described in Sect. 3.2) that genes which only have quantitative effects on fertility escape their detection. This in particularly true for those genes which interact with spermatogenesis without dramatic effects on fertility. The failure to detect the Ste-related site near kl-2 which controls male meiosis is a good example. With most of the X and Y chromosomes used in the screens for male-sterile mutations the males remain fertile when this locus is deleted; therefore, this site was not detected in all the screens which used complete sterility of the males as a criterium. There is no need to stress that all currently used screens miss those genes which have functional copies on the X chromosome and/ or the autosomes.

It has already been mentioned that the structural organization of the genome is crucial for normal spermatogenesis since the majority of X-autosome and Y-autosome translocations is sterile. The parameters of this sterility are only insufficiently known and there is a need for a general rule which defines the structural necessities for the fertility/sterility of those translocations. The elucidation of the sterility of these translocations with the Lifschytz-Lindsley model of X inactivation during spermatogenesis was attempted. However, this model lacks experimental confirmation in *Drosophila*, since cytology strongly argues against an X inactivation during the primary spermatocyte stage (Cooper 1950; Kremer et al. 1986). Therefore, a new theory is needed which considers the activity of the X chromosome during spermatogenesis.

In conclusion, spermatogenesis is a well-studied process in both *D. melanogaster* and *D. hydei*. The information obtained in both species is largely complementary and not redundant. The genetic control of spermatogenesis and the structural genes involved in this process have still to be studied more intensively. A number of genes which are involved in the control of spermatogenesis, the Y-chromosome fertility factors, have quite unusual characteristics. Their study will surely contribute to the demonstration that these genes have functions that are not common to "normal" genes. The crucial function of these genes in spermatogenesis is a challenge for future investigations.

Acknowledgments. I am greatly indebted to M. A. Handel for the critical reading of the manuscript and the revision of the English phrasing. I express my gratitude to W. Hennig, J. A. Kennison, T. C. Kaufman, R. W. Hardy, M. Gatti, and S. Pimpinelli for their generous permission to use material published earlier. R. Brand, W. Hennig, R. Hochstenbach, and D.-H. Lankenau I thank for valuable comments and discussions, M. Smits for typing the manuscript, and R. Dijkhof for the photographic reproductions.

References

Anderson WA (1967) Cytodifferentiation of spermatozoa in *Drosophila melanogaster:* the effect of elevated temperature on spermiogenesis. Mol Gen Genet 99:257–273

Andrews RM, Williamson JH (1975) On nature of Y chromosome fragments in *Drosophila melanogaster* females. III. C(1)RA vs C(1)RM females. Mutat Res 33:213–220

Appels R, Peacock WJ (1978) The arrangement and evolution of highly repeated (satellite) DNA sequences with special reference to *Drosophila*. Int Rev Cytol Suppl 8:69–126

Awgulewitsch A, Wlaschek M, Bünemann H (1985) Different families of repetitive DNA on the Y chromosome of *Drosophila hydei*. Hoppe-Seyler's Z Biol Chem 366:113

Ayles GB, Sanders TG, Kiefer BI, Suzuki DT (1973) Temperature-sensitive mutations in *Drosophila melanogaster*. XI. Male sterile mutants of the Y chromosome. Dev Biol 32:239–257

Baccetti B, Bairati A (1964) Indagini comparative sull'ultrastructura delle cellule germinale maschili in *Dacus oleae* Gmel. ed in *Drosophila melanogaster* Meig (Insecta: Diptera). Redia 49:1–29

Bairati A (1967) Struttura ed ultrastruttura dell' apparato genitale maschile di *Drosophila melanogaster* Meigen. I. Il testiculo. Z Zellforsch Mikrosk Anat 76:56–99

Bairati A, Baccetti B (1965) Indagini comparative sull' ultrastruttura delle cellule germinale maschili in Dacus olea Gmel. ed in *Drosophila melanogaster* Meigen (Insecta: Diptera). II. Nuovi reporti ultrastrutturali sul filamento assile degli spermatozoi. Redia 49:81–85

Bairati A Jr, Baccetti B (1966) Observations on the ultrastructure of male germinal cells in the $X^{Lc}Y^s$ mutant of *Drosophila melanogaster* Meig. Drosophila Inf Serv 41:152

Baker BS, Hall JC (1976) Meiotic mutants: genetic control of meiotic recombination and chromosome segregation. In: Ashburner M, Novitski E (eds) The genetics and biology of *Drosophila*, vol 1 a. Academic Press, London, pp 351–434

Beck H (1976) New compound (1) chromosomes and the production of large quantities of X/O males in *Drosophila hydei*. Genet Res Camb 26:313–317

Beck H, Srdić Ž (1979) Heterochromatin in mitotic chromosomes of *Drosophila hydei*. Genetica (The Hague) 50:1–10

Beermann W, Hess O, Meyer GF (1967) Lampbrush Y chromosomes in spermiogenesis of *Drosophila*. In: Wolff E (ed) The relationship between experimental embryology and molecular biology. Gordon & Breach Science, New York, pp 61–81

Bonaccorsi S, Hackstein J (1987) Cytogenetic dissection of the Y chromosome of *Drosophila hydei*. (In prep)

Bonaccorsi S, Pimpinelli S, Gatti M (1981) Cytological dissection of sex chromosome heterochromatin of *Drosophila hydei*. Chromosoma (Berl) 84:391–403

Brand RC, Hennig W (1987) Testis specific transcription of a poly(A)$^+$ RNA species occurs in Y chromosomal and other genomic sites in *Drosophila hydei* (Submitted for publication)

Breugel FMA van (1970) An analysis of white-mottled mutants in *Drosophila hydei* with observations on X-Y exchanges in the male. Genetica 41:589–625

Breugel FMA van (1971) X-Y exchanges in males of *Drosophila hydei* carrying the w^{m}Co duplication. Genetica (The Hague) 42:1–12

Breugel FMA van, Zijll Langhout B van (1983) The Notch locus of *Drosophila hydei:* alleles, phenotypes and functional organization. Genetics 103:197–217

Bridges CB (1916) Non-disjunction as proof of the chromosome theory of heredity. Genetics 1:1–52 and 107–163

Brosseau GE Jr (1958) Crossing over between Y chromosomes in male *Drosophila*. Drosophila Inf Serv 32:115–116

Brosseau GE Jr (1960) Genetic analysis of the male fertility factors on the Y chromosome of *Drosophila melanogaster*. Genetics 45:257–274

Brosseau GE Jr (1964) Non-randomness in the recovery of detachments from the reversed metacentric compound X chromosome in *Drosophila melanogaster*. Can J Genet Cytol 6:201–206

Brosseau GF Jr, Lindsley DL (1958) A dominantly marked Y chromosome: Y B^s. Drosophila Inf Serv 32:116

Bünemann H (1982) Immobilization of denatured DNA to macroporous supports. II. Steric and kinetic parameters of heterogeneous hybridization reactions. Nucl Acids Res 10:7181–7196

Bünemann H, Westhoff P, Herrmann RG (1982) Immobilization of denatured DNA to macroporous supports. I. Efficiency of different coupling procedures. Nucl Acids Res 10:7163–7180

Callan HG (1986) Lampbrush chromosomes. Springer, Berlin Heidelberg New York

Cooper KW (1950) Normal spermatogenesis in *Drosophila*. In: Demerec M (ed) The biology of *Drosophila*. Hafner, New Haven, Connecticut, pp 1–61

Cooper KW (1956) Phenotypic effects of Y chromosome hyperploidy in *Drosophila melanogaster* and their relation to variegation. Genetics 41:242–264

Cooper KW (1959) Cytogenetic analysis of major heterochromatic elements (especially Xh and Y) in *Drosophila melanogaster* and the theorie of "heterochromatin". Chromosoma (Berl) 10:535–588

Cooper KW (1964) Meiotic conjunctive elements not involving chiasmata. Proc Natl Acad Sci USA 52:1248–1255

Fowler GL (1973a) In vitro cell differentiation in the testes of *Drosophila hydei.* Cell Differ 2:33–41
Fowler GL (1973b) Some aspects of the reproductive biology of *Drosophila:* sperm transfer, sperm storage, and sperm utilization. Adv Genet 17:293–360
Fuller MT (1986) Genetic analysis of spermatogenesis in *Drosophila:* the role of the testis-specific β-tubulin and interacting genes in cellular morphogenesis. In: Gall JG (ed) Gametogenesis and the early embryo. 44th Symp Soc Dev Biol Alan R Liss, New York, pp 19–42
Gatti M, Pimpinelli S (1983) Cytological and genetic analysis of the Y chromosome of *Drosophila melanogaster.* I. Organization of the fertility factors. Chromosoma (Berl) 88:349–373
Gatti M, Pimpinelli S, Santini G (1976) Characterization of *Drosophila* heterochromatin. I. Staining and decondensation with Hoechst 33258 and quinacrine. Chromosoma (Berl) 57:351–375
Glätzer KH (1975) Visualization of gene transcription in spermatocytes of *Drosophila hydei.* Chromosoma (Berl) 53:371–379
Glätzer KH (1984) Preservation of nuclear RNP antigens in male germ cell development of *Drosophila hydei.* Mol Gen Genet 196:236–243
Glätzer KH, Kloetzel PM (1985) Antigens of cytoplasmic RNP particles of D. melanogaster can be localized on distinct Y chromosomal structures in spermatocytes of *D. hydei.* Drosophila Inf Serv 61:84–85
Glätzer KH, Meyer GF (1981) Morphological aspects of the genetic activity in primary spermatocyte nuclei of *Drosophila hydei.* Biol Cell 41:165–172
Goldstein LSB, Hardy RW, Lindsley DL (1982) Structural genes on the Y chromosome of *Drosophila melanogaster.* Proc Natl Acad Sci USA 79:7405–7409
Gould-Somero M, Holland L (1974) The timing of RNA synthesis for spermiogenesis in organ cultures of *Drosophila melanogaster* testes. Wilhelm Roux's Arch Dev Biol 174:133–148
Grell RF (1969) Sterility, lethality, and segregation ratios in XYY males of *Drosophila melanogaster.* Genetics 61:s23–s24
Grond CJ (1984) Spermatogenesis in *Drosophila hydei.* Thesis University of Nijmegen, The Netherlands
Grond CJ, Siegmund I, Hennig W (1983) Visualization of a lampbrush loop-forming fertility gene in *Drosophila hydei.* Chromosome (Berl) 88:50–56
Grond CJ, Rutten RGJ, Hennig W (1984) Ultrastructure of the Y chromosomal lampbrush loops in primary spermatocytes of *Drosophila hydei.* Chromosoma (Berl) 89:85–95
Grond CJ, Kremer H, Hennig W, Kühtreiber W, Freriksen A (1987) Spermiogenesis in *Drosophila hydei.* Wilhelm Roux's Arch Dev Biol (submitted)
Hackstein JHP (1985) Genetic fine structure of the "Th"-"Ps" region of the Y chromosome of *Drosophila hydei.* Hoppe-Seyler's Z Biol Chem 306:117
Hackstein JHP, Leoncini O, Beck H, Peelen G, Hennig W (1982) Genetic fine structure of the Y chromosome of *Drosophila hydei.* Genetics 101:257–277
Hackstein JHP, Hennig W, Steinmann-Zwicky M (1987a) Autosomal control of lampbrush-loop formation during spermatogenesis in *Drosophila hydei* by a gene also affecting somatic sex determination. Wilhelm Roux's Arch Dev Biol 196:119–123
Hackstein JHP, Hennig W, Siegmund I (1987b) Y chromosome-specific mutations induced by a giant transposon in *Drosophila hydei.* Mol Gen Genet (in press)
Hardy RW (1980) Crystal aggregates in the primary spermatocytes of XO males in *Drosophila melanogaster.* Drosophila Inf Serv 55:54–55
Hardy RW, Tokuyasu KT, Lindsley DL (1981) Analysis of spermatogenesis in *Drosophila melanogaster* bearing deletions for Y chromosome fertility genes. Chromosoma (Berl) 83:593–617
Hardy RW, Lindsley DL, Livak KJ, Lewis B, Silverstein AL, Joslyn GL, Edwards J, Bonaccorsi S (1984) Cytogenetic analysis of a segment of the Y chromosome of *Drosophila melanogaster.* Genetics 107:591–610
Hareven D, Zuckerman M, Lifschytz E (1986) Origin and evolution of the transcribed repeated sequences of the Y chromosome lampbrush loops of *Drosophila hydei.* Proc Natl Acad Sci USA 83:125–129
Hartl DL, Hiraizumi Y (1976) Segregation distortion. In: Ashburner M, Novitski E (eds) The genetics and biology of *Drosophila,* vol 1b. Academic Press, London, pp 615–666

Hauschteck-Jungen E, Hartl DL (1978) DNA distribution in spermatid nuclei of normal and SD males of *Drosophila melanogaster*. Genetics 89:15–35

Hauschteck-Jungen E, Hartl DL (1982) Defective histone transition during spermiogenesis in heterozygous segregation distorter males of *Drosophila melanogaster*. Genetics 101:57–69

Hazelrigg T, Fornili P, Kaufman TC (1982) A cytogenetic analysis of X-ray induced male steriles on the Y chromosome of *Drosophila melanogaster*. Chromosoma (Berl) 87:535–559

Heitz E (1933) Cytologische Untersuchungen an Dipteren. III. Die somatische Heteropyknose bei *Drosophila melanogaster* und ihre genetische Bedeutung. Z Zellforsch Mikrosk Anat 20:237–287

Hennig I (1978) Vergleichend-cytologische und -genetische Untersuchungen am Genom der Fruchtfliegen-Arten *Drosophila hydei, neohydei* und *eohydei* (Diptera: Drosophilidae). Entomol Ger 4:211–223

Hennig I (1982) „Hybrid" X-Y translocation chromosomes of *Drosophila hydei* and *D. neohydei*. Chromosome (Berl) 86:491–508

Hennig W (1967) Untersuchungen zur Struktur und Funktion des Lampenbürsten-Y-Chromosoms in der Spermatogenese von *Drosophila*. Chromosoma (Berl) 22:294–357

Hennig W (1972a) Highly repetitive DNA sequences in the genome of *Drosophila hydei*. I. Preferential localization in the Y chromosome heterochromatin. J Mol Biol 71:407–417

Hennig W (1972b) Highly repetitive DNA sequences in the genome of *Drosophila hydei*. II. Occurrence in polytene tissues. J Mol Biol 71:419–431

Hennig W (1977) Gene interactions in germ cell differentiation in *Drosophila*. Adv Enzyme Regul 15:363–371

Hennig W (1978) The lampbrush Y chromosome of the fruit fly species *Drosophila hydei* (Diptera: Drosophilidae). Entomol Ger 4:200–210

Hennig W (1985a) Y chromosome function and spermatogenesis in *Drosophila hydei*. Adv Genet 23:179–234

Hennig W (1985b) The Y chromosome as model system. Hoppe-Seyler's Z Biol Chem 366:118–119

Hennig W (1986) Heterochromatin and germ line-restricted DNA. In: Hennig W (ed) Germ line – soma differentiation. Results and problems in cell differentiation, vol 13. Springer, Berlin Heidelberg New York, pp 175–192

Hennig W, Meyer GF, Hennig I, Leoncini O (1974a) Structure and function of the Y chromosome of *Drosophila hydei*. Cold Spring Harbor Symp Quant Biol 38:673–683

Hennig W, Hennig I, Leoncini O (1974b) Some observations on spermatogenesis of *Drosophila hydei*. Drosophila Inf Serv 51:127

Hennig W, Link B, Leoncini O (1975) The location of the nucleolus organizer region in *Drosophila hydei*. Chromosoma (Berl) 51:57–63

Hennig W, Huijser P, Vogt P, Jäckle H, Edström J-E (1983) Molecular cloning of microdissected lampbrush loop DNA sequences of *Drosophila hydei*. EMBO J 2:1741–1746

Hennig W, Brand RC, Hackstein J, Huijser P, Kirchhoff C, Kremer H, Lankenau DH, Vogt P (1987) Structure and function of Y chromosomal genes in *Drosophila*. Chromosomes Today 9: (in press)

Hess O (1965a) Strukturdifferenzierungen im Y Chromosom von *Drosophila hydei* und ihre Beziehungen zu Gen-Aktivitäten. I. Mutanten der Funktionsstrukturen. Verh Dtsch Zool Ges 28:156–163

Hess O (1965b) Strukturdifferenzierungen im Y-Chromosom von *Drosophila hydei* und ihre Beziehungen zu Gen-Aktivitäten. III. Sequenz und Lokalisation der Schleifenbildungsorte. Chromosoma (Berl) 16:222–248

Hess O (1967a) Complementation of genetic activity in translocated fragments of the Y chromosome in *Drosophila hydei*. Genetics 56:283–295

Hess O (1967b) Morphologische Variabilität der chromosomalen Funktionsstrukturen in den Spermatocytenkernen von *Drosophila*-Arten. Chromosoma (Berl) 21:429–445

Hess O (1968) The function of the lampbrush loops formed by the Y chromosome of *Drosophila hydei* in spermatocyte nuclei. Mol Gen Genet 103:58–71

Hess O (1970) Genetic function correlated with unfolding of lampbrush loops by the Y chromosome in spermatocytes of *Drosophila hydei*. Mol Gen Genet 106:328–346

Hess O (1980) Lampbrush chromosomes. In: Ashburner M, Wright TRF (eds) The genetics and biology of *Drosophila*. Vol 2d. Academic Press, London, pp 1–32

Hess O, Meyer GF (1963a) Artspezifische funktionelle Differenzierungen des Y-Heterochromatins bei *Drosophila*-Arten der *D. hydei*-Subgruppe. Port Acta Biol Ser A VII, 1–2:29–46

Hess O, Meyer GF (1963b) Chromosomal differentiations of the lampbrush type formed by the Y chromosome in *Drosophila hydei* and *D. neohydei*. J Cell Biol 16:527–539

Hess O, Meyer GF (1968) Genetic activities of the Y chromosome in *Drosophila* during spermatogenesis. Adv Genet 14:171–223

Hilliker AJ, Appels R (1982) Pleiotropic effects associated with the deletion of heterochromatin surrounding rDNA on the X chromosome of *Drosophila*. Chromosoma (Berl) 86:469–490

Holmquist G (1975) Hoechst 33258 fluorescent staining of *Drosophila* chromosomes. Chromosoma (Berl) 49:333–356

Huijser P (1985) Molecular structure of the Y chromosome of *Drosophila hydei*. Hoppe-Seyler's Z Biol Chem 366:119

Huijser P, Hennig W (1987) Ribosomal DNA-related sequences in a Y chromosomal lampbrush loop of *Drosophila hydei*. Mol Gen Genet (in press)

Hulsebos T, Hackstein J, Hennig W (1982) Regulation of structural sperm protein synthesis by Y chromosomal loci. In: Jaenicke L (ed) Biochemistry of differentiation and morphogenesis. Springer, Berlin Heidelberg New York, pp 184–188 (Mosbacher Colloqium 33)

Hulsebos TJM, Hackstein JHP, Hennig W (1983) Involvement of Y chromosomal loci in the synthesis of *Drosophila hydei* sperm proteins. Dev Biol 100:238–243

Hulsebos TJM, Hackstein JHP, Hennig W (1984) Lampbrush loopspecific protein of *Drosophila hydei*. Proc Natl Acad Sci USA 81:3404–3408

Ingman-Baker J, Candido EPM (1980) Proteins of the *Drosophila melanogaster* male reproductive system: two-dimensional gel patterns of protein synthesized in the X/O, X/Y and X/Y/Y testis and paragonial gland and evidence that the Y chromosome does not code for structural sperm protein. Biochem Genet 8:809–828

Jürgens G, Wieschaus E, Nüsslein-Volhard C, Kluding H (1984) Mutations affecting the pattern of the larval cuticle in *Drosophila melanogaster*. II. Zygotic loci on the third chromosome. Wilhelm Roux's Arch Dev Biol 193:283–295

Kemphues KJ, Kaufman TC (1980) Two-dimensional gel analysis of total proteins from X/O, X/Y, X/Y/Y, X/Y^S and X/Y^L testes from *D. melanogaster*. Drosophila Inf Serv 55:72

Kemphues KJ, Raff RA, Kaufman TC, Raff EC (1979) Mutation in a structural gene for a β-tubulin specific to testis in *Drosophila melanogaster*. Proc Natl Acad Sci USA 76:3991–3995

Kemphues KJ, Raff EC, Kaufman TC (1983) Genetic analysis of B2t, the structural gene for a testis-specific β-tubulin subunit in *Drosophila melanogaster*. Genetics 105:345–356

Kennison JA (1981) The genetic and cytological organization of the Y chromosome of *Drosophila melanogaster*. Genetics 98:529–548

Kennison JA (1983) Analysis of Y-linked mutations to male sterility in *Drosophila melanogaster*. Genetics 103:219–234

Kiefer BI (1966) Ultrastructural abnormalities in developing sperm of X/O *Drosophila melanogaster*. Genetics 54:1441–1452

Kiefer BI (1968) Y-mutants and spermiogenesis in *Drosophila melanogaster*. Genetics 60:192

Kiefer BI (1969) Phenotypic effects of Y chromosome mutations in *Drosophila melanogaster*. I. Spermiogenesis and sterility in kl-1$^-$ males. Genetics 61:157–166

Kiefer BI (1970) Development, organization, and degeneration of the *Drosophila* sperm flagellum. J Cell Sci 6:177–194

Kiefer BI (1973) Genetics of sperm development in *Drosophila*. In: Ruddle EH (ed) Genetic mechanisms in development. Academic Press, London, pp 47–102

Kloetzel PM, Knust E, Schwochau M (1981) Analysis of nuclear proteins in primary spermatocytes of *Drosophila hydei:* the correlation of nuclear proteins with the function of the Y chromosome loops. Chromosoma (Berl) 84:67–86

Kremer H, Hennig W, Dijkhof R (1986) Chromatin organization in the male germ line of *Drosophila hydei*. Chromosoma (Berl) 94:147–161

Lehmann R, Jimènez F, Dietrich U, Campos-Ortega JA (1983) On the phenotype of early neurogenesis in *Drosophila melanogaster*. Wilhelm Roux's Arch Dev Biol 192:67–74

Leoncini O (1977) Temperatursensitive Mutanten im Y-Chromosom von *Drosophila hydei*. Chromosoma (Berl) 63:329–357

Liebrich W (1981 a) In vitro differentiation of isolated single spermatocyte cysts of *Drosophila hydei*. Eur J Cell Biol 24:335
Liebrich W (1981 b) In vitro spermatogenesis in *Drosophila*. I. Development of isolated spermatocyte cysts from wild-type *D. hydei*. Cell Tissue Res 220:251–262
Lifschytz E (1972) X-Chromosome inactivation: an essential feature of normal spermiogenesis in male heterogametic organisms. In: Beatty RA, Gluecksohn-Waelsch S (eds) The genetics of the spermatozoon. Proc Int Symp Edinburgh, pp 223–232
Lifschytz E (1974) Genes controlling chromosome activity. An X-linked mutation affecting Y lampbrush loop activity in *Drosophila hydei*. Chromosoma (Berl) 47:415–427
Lifschytz E (1975) Genes controlling chromosome activity. The role of genes blocking Y-lampbrush loop propagation. Chromosoma (Berl) 53:231–241
Lifschytz E (1979) A procedure for the cloning and identification of Y-specific middle repetitive sequences in *Drosophila hydei*. J Mol Biol 133:267–277
Lifschytz E, Hareven D (1977) Gene expression and the control of spermatid morphogenesis in *Drosophila melanogaster*. Dev Biol 58:276–294
Lifschytz E, Lindsley DL (1972) The role of X-chromosome inactivation during spermatogenesis. Proc Natl Acad Sci USA 69:182–186
Lifschytz E, Meyer GF (1977) Characterization of male meiotic sterile mutations in *Drosophila melanogaster*. The genetic control of meiotic divisions and gametogenesis. Chromosoma (Berl) 64:371–392
Lifschytz E, Hareven D, Azriel A, Brodsly H (1983) DNA clones and RNA transcripts of four lampbrush loops from the Y chromosome of *Drosophila hydei*. Cell 32:191–199
Lindsley DL (1955) Spermatogonial exchange between the X and Y chromosome of *Drosophila melanogaster*. Genetics 40:24–44
Lindsley DL (1982) The genetics of male fertility. Drosophila Inf Serv 58:2
Lindsley DL, Grell EH (1969) Spermiogenesis without chromosomes in *Drosophila melanogaster*. Genetics 61 Suppl (1):69–78
Lindsley DL, Lifschytz E (1972) The genetic control of spermatogenesis in *Drosophila*. In: Beatty RA, Gluecksohn-Waelsch S (eds) The genetics of the spermatozoon. Proc Int Symp Edinburgh, pp 203–222
Lindsley DL, Sandler L (1958) The meiotic behaviour of grossly deleted X chromosomes in *Drosophila melanogaster*. Genetics 43:547–563
Lindsley DL, Tokuyasu KT (1980) Spermatogenesis. In: Ashburner M, Wright TRF (eds) The genetics and biology of *Drosophila*, vol 2d. Academic Press, London, pp 225–294
Lindsley DL, Sandler L, Baker BS, Carpenter ATC, Denell RE, Hall JC, Jacobs PA, Miklos GLG, Davis BK, Gethmann RC, Hardy RW, Hessler A, Miller SM, Nozawa H, Parry DM, Gould-Somero M (1972) Segmental aneuploidy and the genetic structure of the *Drosophila* genome. Genetics 71:157–184
Lindsley DL, Pearson C, Rokop SR, Jones M, Stern D (1979) Genotypic features causing sterility of males carrying a bobbed-deficient X chromosome: translocations involving chromosome 2 and 3. Genetics 91:s69–s70
Livak KJ (1984) Organization and mapping of a sequence on the *Drosophila melanogaster* X and Y chromosomes that is transcribed during spermatogenesis. Genetics 107:611–634
Loos F de, Dijkhof R, Grond CJ, Hennig W (1984) Lampbrush loopspecificity of transcript morphology in spermatocyte nuclei of *Drosophila hydei*. EMBO J 3:2845–2849
Lovett J (1983) Molecular aspects of Y chromosome function in *Drosophila melanogaster* spermiogenesis. Ph D Thesis Indiana University, Bloomington, Indiana, USA
Lovett J, Kaufman TC, Mahowald AP (1980) A locus on the X chromosome apparently controlled by the Y chromosome during spermatogenesis in *Drosophila melanogaster*. Eur J Cell Biol 22:49
Lucchesi JC (1965) The nature of induced exchanges between the attached-X and Y chromosomes of *Drosophila melanogaster* females. Genetics 51:209–216
Lyon MF (1961) Gene action in the X chromosome of the mouse. Nature 190:372–373
Melzer S, Glätzer KH (1985) Localization of RNP antigens in primary spermatocytes of *Drosophila melanogaster* by indirect immunofluorescence and their correlation to fertility factors. Drosophila Inf Serv 61:121–122

Merriam JR (1968) The meiotic behaviour of tandem acrocentric compound X chromosomes in *Drosophila melanogaster*. Genetics 59:361–366

Meyer GF (1961) Interzelluläre Brücken im Hoden und im Ei-Nährzellverband von *Drosophila melanogaster*. Z Zellforsch Mikrosk Anat 54:238–251

Meyer GF (1963) Die Funktionsstrukturen des Y-Chromosoms in den Spermatocytenkernen von *Drosophila hydei*, *D. neohydei*, *D. repleta* und einigen anderen *Drosophila*-Arten. Chromosoma (Berl) 14:207–255

Meyer GF (1964) Die parakristallinen Körper in den Spermienschwänzen von *Drosophila*. Z Zellforsch Mikrosk Anat 62:762–784

Meyer GF (1968) Spermiogenese in normalen und Y-defizienten Männchen von *Drosophila melanogaster* und *D. hydei*. Z Zellforsch Mikrosk Anat 84:141–175

Meyer GF (1972) Influence of Y chromosome on fertility and phenotype of *Drosophila* spermatozoa. In: Beatty RA, Gluecksohn-Waelsch S (eds) The genetics of the spermatozoon. Proc Int Symp Edinburgh, pp 387–405

Meyer GF, Hennig W (1974a) The nucleolus in primary spermatocytes of *Drosophila hydei*. Chromosoma (Berl) 46:121–144

Meyer GF, Hennig W (1974b) Molecular aspects of the fertility factors in *Drosophila*. In: Afzelius BA (ed) The functional anatomy of the spermatozoon. Pergamon, Oxford, New York, pp 69–75

Meyer GF, Hess O, Beermann W (1961) Phasenspezifische Funktionsstrukturen in Spermatocytenkernen von *Drosophila melanogaster* und ihre Abhängigkeit vom Y-Chromosom. Chromosoma (Berl) 12:676–716

Miller OL Jr, Beatty BR (1969) Visualization of nucleolar genes. Science 164:955–957

Neuhaus M (1936) Crossing over between X- and Y-chromosomes in the female of *Drosophila melanogaster*. Z Indukt Abstammungs-Vererbungsl 71:265–275

Neuhaus M (1937) Additional data on crossing-over between X- and Y-chromosomes in *Drosophila melanogaster*. Genetics 22:333–339

Neuhaus MJ (1939) A cytogenetic study of the Y chromosome of *Drosophila melanogaster*. J Genet 37:229–254

Nicoletti B, Lindsley DL (1960) Translocations between the X and the Y chromosomes of *Drosophila melanogaster*. Genetics 45:1705–1722

Nüsslein-Volhard C, Wieschaus E, Kluding H (1984) Mutations affecting the pattern of the larval cuticle in *Drosophila melanogaster*. I. Zygotic loci on the second chromosome. Wilhelm Roux's Arch Dev Biol 193:267–282

Olivieri G, Olivieri A (1965) Autoradiographic study of nucleic acid synthesis during spermatogenesis in *Drosophila melanogaster*. Mutat Res 2:366–380

Parker DR (1967) Induced heterologous exchange at meiosis in *Drosophila*. I. Exchanges between Y and fourth chromosomes. Mutat Res 4:333–337

Parker DR, Hammond AE (1958) The production of translocations in *Drosophila* oocytes. Genetics 43:92–100

Peacock WJ, Miklos GLG (1973) Meiotic drive in *Drosophila*: new interpretations of the segregation distorter and sex chromosome systems. Adv Genet 17:361–409

Peacock WJ, Lohe AR, Gerlach WL, Dunsmuir P, Dennis ES, Appels R (1978) Fine structure and evolution of DNA in heterochromatin. Cold Spring Harbor Symp Quant Biol 42:1121–1135

Perotti ME (1969) Ultrastructure of the mature sperm of *Drosophila melanogaster* Meig. J Submicrosc Cytol 1:171–196

Pimpinelli S, Santini G, Gatti M (1976) Characterization of *Drosophila* heterochromatin. II. C- and N-banding. Chromosoma (Berl) 57:377–386

Renkawitz R (1978a) Characterization of two moderately repetitive DNA components within the β-heterochromatin of *Drosophila hydei*. Chromosoma (Berl) 66:225–236

Renkawitz R (1978b) Two highly repetitive DNA satellites of *Drosophila hydei* localized within the α-heterochromatin of specific chromosomes. Chromosoma (Berl) 66:237–248

Renkawitz R (1979) Isolation of twelve satellite DNAs from *Drosophila hydei*. Int J Biol Macromol 1:133–136

Ritossa F (1976) The bobbed locus. In: Ashburner M, Novitski E (eds) The genetics and biology of *Drosophila*, vol 1 b. Academic Press, London, pp 801–846

Ritossa F, Spiegelman S (1965) Localization of DNA complementary to ribosomal RNA in the nucleolus organizer region of Drosophila melanogaster. Genetics 53:737–745

Safir S (1920) Genetic and cytological examination of the phenomena of primary nondisjunction in Drosophila melanogaster. Genetics 5:459–487

Schäfer U (1978) Sterility in Drosophila hydei × D. neohydei hybrids. Genetica 49:205–214

Schäfer U (1979) Viability in Drosophila hydei × D. neohydei hybrids and its regulation by genes located in the sex heterochromatin. Biol Zentralbl 98:153–161

Shellenbarger DL, Cross DP (1977) A new class of male-sterile mutations with combined temperature-sensitive lethal effects in Drosophila melanogaster. Genetics 86:s358

Shen TH (1932) Cytologische Untersuchungen über Sterilität bei Männchen von Drosophila melanogaster und bei F_1-Männchen der Kreuzung zwischen Drosophila simulans-Weibchen und Drosophila melanogaster-Männchen. Z Zellforsch Mikrosk Anat 15:547–580

Shoup JR (1967) Spermiogenesis in wild type and in a male sterility mutant of Drosophila melanogaster. J Cell Biol 32:663–675

Spofford JB (1976) Position-effect variegation in Drosophila. In: Ashburner M, Novitski E (eds) The genetics and biology of Drosophila, vol 1c. Academic Press, London, pp 955–1018

Steffensen DL, Appels R, Peacock WJ (1981) The distribution of two highly repeated DNA sequences within Drosophila melanogaster chromosomes. Chromosoma (Berl) 82:525–541

Stern C (1927) Ein genetischer und zytologischer Beweis für Vererbung im Y-Chromosom von Drosophila melanogaster. Z Indukt Abstammungs-Vererbungsl 44:187–231

Stern C (1929) Untersuchungen über Aberrationen des Y-Chromosoms von Drosophila melanogaster. Z Indukt Abstammungs-Vererbungsl 51:253–353

Tates AD (1971) Cytodifferentiation during spermatogenesis in Drosophila melanogaster. Thesis University of Leiden, The Netherlands

Tokuyasu KT (1974a) Dynamics of spermiogenesis in Drosophila melanogaster. III. Relation between axoneme and mitochondrial derivatives. Exp Cell Res 84:239–250

Tokuyasu KT (1974b) Dynamics of spermiogenesis in Drosophila melanogaster. IV. Nuclear transformation. J Ultrastruct Res 48:284–303

Tokuyasu KT (1974c) Spoke heads in sperm tail of Drosophila melanogaster. J Cell Biol 63:334–337

Tokuyasu KT (1975a) Dynamics of spermiogenesis in Drosophila melanogaster. V. Head tail alignment. J Ultrastruct Res 50:117–129

Tokuyasu KT (1975b) Dynamics of spermiogenesis in Drosophila melanogaster. VI. Significance of "onion" nebenkern formation. J Ultrastruct Res 53:93–112

Tokuyasu KT, Peacock WJ, Hardy RW (1972a) Dynamics of spermiogenesis in Drosophila melanogaster. I. Individualization process. Z Zellforsch Mikrosk Anat 124:479–506

Tokuyasu KT, Peacock WJ, Hardy RW (1972b) Dynamics of spermiogenesis in Drosophila melanogaster. II. Coiling process. Z Zellforsch Mikrosk Anat 127:492–525

Tokuyasu KT, Peacock WJ, Hardy RW (1977) Dynamics of spermiogenesis in Drosophila melanogaster. VII. Affects of Segregation Distorter (SD) chromosome. J Ultrastruct Res 58:96–107

Vogt P (1985) Molecular structure of the lampbrush loop nooses on the Y chromosome of Drosophila hydei. Hoppe-Seyler's Z Biol Chem 366:125

Vogt P, Hennig W (1983) Y chromosomal DNA of Drosophila hydei. J Mol Biol 167:37–56

Vogt P, Hennig W (1986a) Molecular structure of the lampbrush loop nooses of the Y chromosome of Drosophila hydei. I. The Y chromosome-specific repetitive DNA sequence family ayl is dispersed in the loop DNA. Chromosoma (Berl) 94:449–458

Vogt P, Hennig W (1986b) Molecular structure of the lampbrush loop nooses of the Y chromosome of Drosophila hydei. II. DNA sequences with homologies to multiple genomic locations are a major constituent of the loop. Chromosoma (Berl) 94:459–467

Vogt P, Hennig W, Siegmund I (1982) Identification of cloned Y chromosomal DNA sequences from a lampbrush loop of Drosophila hydei. Proc Natl Acad Sci USA 79:5132–5136

Vogt P, Hennig W, ten Hacken D, Verbost P (1986) Evolution of Y chromosomal lampbrush loop DNA sequences of Drosophila. Chromosoma (Berl) 94:367–376

Wieschaus E, Nüsslein-Volhard C, Jürgens G (1984) Mutations affecting the pattern of the larval cuticle in Drosophila melanogaster. III. Zygotic loci on the X chromosome and the fourth chromosome. Wilhelm Roux's Arch Dev Biol 193:296–307

Williamson JH (1970) Ethyl Methanesulfonate-induced mutants in the Y chromosome of *Drosophila melanogaster*. Mutat Res 10:597–605
Williamson JH (1972) Allelic complementation between mutants in the fertility factors of the Y chromosome in *Drosophila melanogaster*. Mol Gen Genet 119:43–47
Williamson JH (1976) The genetics of the Y chromosome. In: Ashburner M, Novitski E (eds) The genetics and biology of *Drosophila*, vol 1 b. Academic Press, London, pp 667–699
Williamson JH, Meidinger E (1979) Y chromosome hyperploidy and male fertility in *Drosophila melanogaster*. Can J Genet Cytol 21:21–24
Yamasaki N (1977) Selective staining of Y chromosomal loop DNA in *Drosophila hydei, D. neohydei* and *D. eohydei*. Chromosoma (Berl) 60:27–37
Yamasaki N (1981) Differential staining of Y chromosome lampbrush loops of *Drosophila hydei*. Chromosoma (Berl) 83:679–684
Zacharias H, Hennig W, Leoncini O (1982) Microspectrophotometric comparison of the genome sizes of *Drosophila hydei* and some related species. Genetica (The Hague) 58:153–157
Zimmering S (1976) Genetic and cytogenetic aspects of altered segregation phenomena in *Drosophila*. In: Ashburner M, Novitski E (eds) The genetics and biology of *Drosophila*, vol 1 b. Academic Press, London, pp 569–613

Genetic Control of Sex Determination in the Germ Line of *C. elegans*

JUDITH KIMBLE [1]

1 Introduction

How is a germ cell instructed to differentiate as a sperm or an oocyte? This is essentially a question of sex determination asked at the cellular level. Sex determination in the germ line has been elusive to experimentation in both *Drosophila* and in mice. However, in the nematode, *Caenorhabditis elegans,* control over the choice between spermatogenesis and oogenesis has proven to be particularly accessible to genetic analysis. In this chapter, we review our current knowledge of the mechanisms in *C. elegans* that influence this decision.

2 Background

2.1 The Organism, *C. elegans*

C. elegans, a nonparasitic soil nematode, is uniquely suited to genetic analyses of animal development. Its main advantages for such studies are its short life cycle (3 days), ease of maintenance, and simple anatomy (Brenner 1974). The transparency of this tiny worm throughout its development permits the direct examination of individual cells as they divide, migrate, and differentiate in the living animal. The invariant cell lineage of the somatic tissues of *C. elegans* has led to the complete description of the ancestry of all somatic structures (Sulston et al. 1983). Furthermore, the small size of the *C. elegans* genome (about 20 X *E. coli*, about 1/35th man) facilitates both classic mutational analyses and more modern molecular studies of individual genes (Sulston and Brenner 1974).

C. elegans can develop as either of two sexes (Fig. 1). Diploid animals with two X chromosomes (XX) are self-fertilizing hermaphrodites; with only a single X chromosome (XO), they are males. Reproduction in *C. elegans* can occur either by self-fertilization, in which oocytes are fertilized by the hermaphrodite's own sperm, or by cross-fertilization. The two sexes differ substantially in morphology, biochemistry, and behavior. The hermaphrodite is essentially a female that makes a limited number of sperm. Hermaphrodites possess a gonad with two equivalent

---

[1] Department of Biochemistry, College of Agricultural and Life Sciences, and Laboratory of Molecular Biology, Graduate School, University of Wisconsin-Madison, 1525 Linden Drive, Madison, Wisconsin 53706, USA.

Results and Problems in Cell Differentiation, 15
Spermatogenesis: Genetic Aspects
Edited by W. Hennig
© Springer-Verlag Berlin Heidelberg 1987

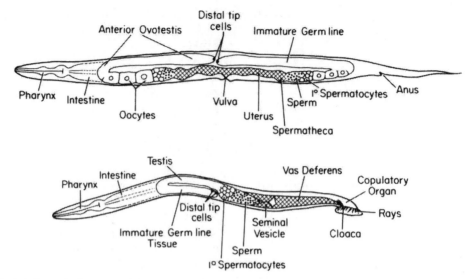

Fig. 1. Schematic diagrams of a young adult hermaphrodite (*above*) and a male (*below*). Notice the sexually dimorphic gonads (two ovotestes in the hermaphrodite, one testis in the male) and tails (a whiplike tail in the hermaphrodite, a tail specialized for copulation in the male). The somatic gonad of each is indicated by *cross-hatching*, the rest of the tubular gonad is germ line

ovotestes that meet centrally at a vulva (Hirsh et al. 1976). Each ovotestis produces first about 150 sperm, and then oocytes. Males have a gonad with a single testis that opens to the exterior posteriorly in the tail (Klass et al. 1976). The testis produces sperm continuously. During mating, the male clasps the hermaphrodite with its specialized tail and ejaculates sperm into the hermaphrodite uterus through the vulva. The male tail is an elaborate sex-specific structure composed of special nerves, muscles, and hypodermal cells that are required for mating (Sulston et al. 1980). The intestine is biochemically, but not morphologically, different in the two sexes. Only the hermaphrodite intestine produces yolk proteins (Kimble and Sharrock 1983).

Over 600 of the estimated 3000 genes in *C. elegans* have been identified by mutation. These include many genes that control various aspects of development and behavior. In the standardized genetic nomenclature of *C. elegans*, a gene is specified by a three letter code indicating its mutant phenotype and a number, e.g., *fem-1*. Mutant alleles are named by a letter specifying the laboratory of its origin and a number, e.g., *e1*, the first allele isolated by S. Brenner in England, and mutant phenotypes are given a three letter code, e.g., fem for feminized.

Three sophisticated genetic tools have recently become available for analysis of development in *C. elegans*. First, suppressor mutations have been identified that encode amber suppressor tRNAs and promote readthrough of amber (UAG) terminator codons of many genes (Wills et al. 1983). These suppressors can be used for identification of amber alleles at many genetic loci. The existence of an amber allele at a particular locus provides important evidence that the gene encodes a protein. Furthermore, the mutant phenotype of an animal homozygous for an amber allele is often the null phenotype. Second, the generation of animals mosaic for a given activity has been achieved by the discovery that the loss of a free duplication during mitosis can be used to generate such animals (Herman 1984). Such genetic mosaic animals provide information on the cellular or tissue specificity of a gene's function. Third, a transposable element, Tc1, has recently been discovered in *C. elegans* (Moerman and Waterston 1984; Eide and Anderson 1985). Mutations caused by insertion of Tc1 have already been used successfully for cloning genes by transposon tagging (Greenwald 1985).

2.2 Development and Anatomy of the Germ Line

The germ line of *C. elegans* develops as a clone from a single embryonic precursor cell, P_4 (Sulston et al. 1983). P_4 divides once during embryogenesis; its two daughter cells are the germ line progenitor cells that are present at hatching. These two cells are called Z2 and Z3. During postembryonic development, Z2 and Z3 divide mitotically to generate about 2000 nuclei in hermaphrodites and about 1000 nuclei in males. The postembryonic pattern of cell divisions in the germ line is not fixed (Kimble and Hirsh 1979). Thus, no strict cell lineage has been deduced for germ line descendants. This contrasts markedly with the development of somatic tissues that occurs by a rigidly invariant pattern of cell divisions.

The germ line nuclei are located at the periphery of a large syncytial tube, each partially enclosed by a membrane (Hirsh et al. 1976). Some germ line nuclei remain mitotic throughout the life of the animal; these mitotically dividing nuclei are loated at the distal end of the germ line tube and serve the function of stem cells for the germ line. The other germ line nuclei, located more proximally, enter the meiotic cell cycle and mature through the stages of meiosis and gametogenesis as they progress proximally (Klass et al. 1976). This means that the germ line tissue possesses a polarity of maturation with the least mature nuclei at one end (distal end), nuclei in meiotic prophase in the middle, and maturing gametes at the other end (proximal end).

Two somatic cells, the distal tip cells, regulate entry into meiosis from mitosis (Kimble and White 1981). In hermaphrodites, one distal tip cell occupies the distal end of each of the two equivalent ovotestes; in males, both distal tip cells occupy the distal end of the single testis (Fig. 1). The distal tip cells are descendants of the progenitor cells of the somatic gonad, Z1 and Z4 (Kimble and Hirsh 1979). The regulatory role of the distal tip cell was discovered by a series of laser ablation experiments (Kimble and White 1981). When both distal tip cells are destroyed by a laser microbeam, all germ line nuclei enter meiosis after only a few (one to four) mitotic divisions.

The proximal arm of the ovotestis (hermaphrodite) or the testis (male) is the site of gametogenesis. In hermaphrodites, the proximal arm is encapsulated by a somatic contractile epithelial sheath, or oviduct. In males, this arm is only partly ensheathed somatically by the most distal cells of the seminal vesicle. A detailed description of the processes of spermatogenesis and oogenesis is beyond the scope of this chapter. (For a recent review, see Kimble and Ward 1987.) Briefly, each tetraploid primary spermatocyte generates four haploid spermatids. During the meiotic divisions, a large body of the primary spermatocyte's cytoplasm is excluded from the emerging spermatids. The spermatid matures to a spermatozoon by rearrangement of its cellular constituents to form a pseudopod. The mature spermatozoa of *C. elegans* are asymmetrical, amoeboid cells.

Oogenesis begins in hermaphrodites after about 150 sperm have been made in each ovotestis. Once oogenesis has begun, the proximal arm of the ovotestis consists of a single file of enlarging oocytes. Each of these oocytes contains a single nucleus with chromosomes in late meiotic prophase. Prior to fertilization, oocytes are halted at diakinesis of meiotoic prophase I. Both meiotic divisions occur after fertilization. When an oocyte is fertilized, the zygote moves through the spermatheca and into the uterus where meiosis of the oocyte nucleus is completed. Two polar bodies are extruded, the two pronuclei become apposed after a complex series of movements, and embryonic divisions begin.

2.3 The Global Sex Determination Genes

The initial signal for sex determination in *C. elegans* is the ratio of X chromosomes to autosomes (Madl and Herman 1978). This signal appears to regulate the activity of seven genes which, in turn, regulate the sexual phenotype. The mutant phenotypes of these seven genes are summarized in Table 1. A loss-of-function (lf) mutation of any of these seven global sex determination genes has a dramatic effect on the sexual differentiation of most or all tissues of *C. elegans*. Thus, XX animals homozygous for a null mutation in any of the three *tra* genes develop along the male pathway of development, whereas the development of XO animals is unaffected (Hodgkin and Brenner 1977). Since the absence of any of the *tra* products leads to this phenotype, the wild-type function of the three *tra* genes must normally be required for the development of XX animals as hermaphrodites. Conversely, both XX and XO animals homozygous for a loss-of-function mutation in any of three *fem* genes are feminized (Nelson et al. 1978; Kimble et al. 1984; Donaich and Hodgkin 1984; Hodgkin 1986). The feminized XX animal

Table 1. Global sex determination genes in *C. elegans*

Gene	Loss-of-function mutant phenotype			
	XX		XO	
	Soma	Germ line	Soma	Germ line
Wild type	Female	Sperm, oocytes	Male	Sperm
fem-1	Female	Oocytes	Female	Oocytes
fem-2	Female	Oocytes	Female	Oocytes
fem-3	Female	Oocytes	Female	Oocytes
tra-1	Male	Sperm, oocytes	Male	Sperm, oocytes
tra-2	Pseudomale	Sperm	Male	Sperm
tra-3	Pseudomale	Sperm, oocytes	Male	Sperm
her-1	Female	Sperm, oocytes	Female	Sperm, oocytes

no longer makes sperm, but somatically, it is identical to a hermaphrodite XX animal. Thus, based on the same reasoning as used for the *tra* genes, the wild-type gene products of each of the *fem* genes must be necessary for normal male development – both in the somatic tissues and in the germ line. Finally, XO animals homozygous for a recessive mutation of *her-1* develop as hermaphrodites, whereas XX *her-1* animals remain unaffected (Hodgkin 1980). The wild-type *her-1* gene product must therefore be required for male development in XO, but not in XX animals.

Dominant gain-of-function (gf) alleles have been isolated for four of the seven global sex-determining genes described above (Hodgkin 1983a; Trent et al. 1983; Barton et al. 1987; Doniach 1986). In all four cases, the effect on sexual phenotype of a gain-of-function mutation is opposite to the effect of a loss-of-function mutation at the same locus. Thus, for example, a loss-of-function allele of *her-1* "hermaphroditizes" XO animals, whereas a gain-of-function allele of *her-1* masculinizes XX animals (Trent et al. 1983). The finding that gain-of-function and loss-of-function alleles confer opposite phenotypes supports the idea that these genes are central to the determination of the sexual phenotype.

The epistatic relationships observed in strains harboring more than one sex determination mutation have led to a formal model of genetic regulation of sex determination (Hodgkin 1986). Figure 2 shows a simplified version of the proposed genetic pathway as it pertains to control of the sexual phenotype in somatic tissues. This pathway consists of a cascade of negative regulators that ultimately controls the state of activity of a single gene, *tra-1*. Although gene activity is indicated as ON or OFF, this is a genetic formalism and is not meant to imply a molecular mechanism of regulation. The temperature-sensitive periods of the sex determination genes for which temperature-sensitive alleles have been isolated concur with this suggested order (Hodgkin 1983b).

The role of *tra-1* as the primary sex-determining gene in the somatic tissues (Fig. 2) is based on the finding that the somatic mutant phenotype of *tra-1* is epistatic to the mutant phenotypes of all of the other sex-determining genes (Hodgkin 1980). This result suggests that the state of *tra-1* dictates the sexual phenotype

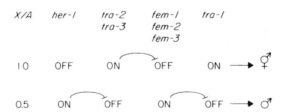

Fig. 2. Simplified pathway of genetic control of sexual determination in the somatic tissues. To achieve hermaphrodite somatic development, *tra-2* and *tra-3* negatively regulate the *fem* genes resulting in activity of *tra-1*, the gene proposed to direct hermaphrodite (female) somatic development. To achieve male development, the ratio of X chromosomes to autosomes triggers the activity of *her-1* to negatively regulate *tra-2* and *tra-3*; therefore, the *fem* genes are active and they, in turn, can negatively regulate *tra-1*. The level at which this negative regulation is executed, e.g., transcriptional, posttranscriptional, is not known. (See Hodgkin 1986 for his most recent model)

of the soma. In the germ line, the primary sex-determining genes appear to be the *fem* genes. Thus, the mutant phenotype of any of the *fem* genes is epistatic to that of the other mutations in the germ line tissue. This means that an animal homozygous for loss-of-function mutations in both *fem-1* and *tra-1*, for example, are male somatically since the *tra-1* phenotype is epistatic there, but it is female in the germ line since the *fem-1* phenotype is epistatic in this tissue (Doniach and Hodgkin 1984). The role of *tra-1* in the germ line is not understood. The null phenotype of *tra-1*, though clear in the soma (Hodgkin 1983a) is elusive in the germ line (Kimble and Schedl 1987). The *tra-1* gene may encode a single function that is specific to somatic tissues, or it may be more complex, encoding multiple products or a product with diverse functions that depend on the site of action.

3 Control of the Sperm/Oocyte Decision

In my laboratory, we have chosen to analyze sex determination in a single tissue – the germ line. Our decision to study a single tissue stems from the differences observed in the effect of mutations of the global sex determination on the sexual phenotype of the germ line vs the somatic tissues. Our decision to focus on the germ line is based on the relative simplicity of the sperm/oocyte decision compared to the formation of complex sex-specific somatic organs, on the genetic selections available for germ line sex determination mutants, and on the ability to microinject the syncytial germ line for bioassays once regulatory molecules have been isolated biochemically (Kimble et al. 1982).

3.1 Influence of Cell Ancestry on the Sperm/Oocyte Decision

Several lines of evidence suggest that the exact lineal descent of a germline cell from its progenitor, P_4, is not critical to its differentiation as either sperm or oocyte. First, in direct contrast to the development of somatic structures, where cell ancestry is critical to a cell's fate, the cell division pattern of the germ line is variable (Kimble and Hirsh 1979). Second, wild-type cells that would normally give rise to oocytes can differentiate as sperm instead (after laser ablation of the distal tip cells, Kimble and White 1981). These experiments indicate that no precursor cell that is destined to generate oocytes is set aside early in development. Third, the kind of gamete produced by the hermaphrodite germ line can be switched from spermatogenesis to oogenesis or from oogenesis to spermatogenesis by a simple temperature shift during adulthood in animals carrying a gain-of-function allele of *fem-3* (Barton et al. 1987). This last result suggests that sex determination in the germ line is a continuing process. Thus, germ line precursor cells are not committed during early development to generate either sperm or oocyte descendants. Indeed, at least some germ line cells remain uncommitted even in the adult animal.

3.2 Influence of Chromosomal Sex on the Sperm/Oocyte Decision

Both sperm and oocytes can be made by either XX or XO cells. In wild-type animals, oocytes are produced by XX cells and sperm are made by either XX or XO cells. In various mutant strains, functional oocytes are made by XO germ cells [e.g., *her-1*, Hodgkin 1980; *tra-1*(gf), Hodgkin 1983a]. Thus, the chromosome sex of a germ cell does not prevent its differentiation as either sperm or oocyte.

3.3 Influence of the Soma on the Sperm/Oocyte Decision

The soma does not seem to be crucial to specification of a germ cell as sperm or oocyte. The wild-type female (hermaphrodite) produces both sperm and oocytes. Certain mutant hermaphrodites can produce only oocytes (see Tables 1 and 2) and others produce only sperm [see Table 2, see also below, *fem-3(gf)*] in a morphologically normal XX hermaphrodite soma. Conversely, certain mutant males (see Table 2) make germ cells that resemble oocytes in a morphologically normal male soma. Such oocyte-like cells are large and blocky in comparison to the tiny amoeboid sperm, and they contain oocyte-like granules.

Furthermore, in both hermaphrodites and males, all somatic structures (e.g., seminal vesicle, uterus) can be deleted by laser ablation of a few somatic precursor cells during L1 without blocking the initiation of sperm or oocyte formation (Kimble and White 1981). The distal tip cells must remain intact, presumably to permit sufficient proliferation of the germ line that makes both sperm and oocytes.

Although somatic structures are not required for the onset of spermatogenesis or oogenesis, hermaphrodite structures are necessary for the normal maturation of a germ cell into a functional oocyte. The oocytes produced either in a hermaphrodite gonad lacking somatic structures, or in a male gonad, are generally smaller and more irregularly shaped than normal. Thus, the hermaphrodite oviduct that surrounds maturing oocytes may be critical to normal oocyte morphogenesis.

Table 2. Germ line-specific sex determination genes in *C. elegans*

Gene	Loss-of-function mutant phenotype	
	XX	XO
Wild type	Sperm, oocytes	Sperm
fog-1	Oocytes	Oocytes
fog-2	Oocytes	Sperm
fog(q126)I	Oocytes	Sperm, oocytes
mog-1	Sperm	Sperm
mog-2	Sperm	Sperm

3.4 Genetics of Sex Determination in the Germ Line Tissue

Germ cells are directed to differentiate as either sperm or oocyte by a small number of control genes. These genes fall into two classes. The global sex determination genes, as described above, influence the specification of the sexual phenotype of all tissues in the animal, including both somatic and germ line tissues. In addition, germ line-specific sex determination genes exist that influence the choice of sexual phenotype in germ cells. The latter class includes *fog* genes, for *f*eminization *of* the *g*erm line, and *mog* genes, for *m*asculinization *of* the *g*erm line. The *fog* and *mog* genes and their mutant phenotypes are summarized in Table 2.

3.4.1 Selections and Screens for Mutations that Sexually Transform the Germ Line

We have devised genetic selections to isolate mutations that sexually transform the germ line (Kimble et al. 1986). The idea behind these genetic selections is simple. Basically, an XX hermaphrodite is self-sterile if it produces only sperm or only oocytes, but it is self-fertile if sperm and then oocytes are made. Therefore, mutants that feminize the hermaphrodite germ line so that only oocytes are made (or masculinize it so that only sperm are made) can be used to select for suppressor mutations that reinstate self-fertility. Selection of both masculinizing and feminizing suppressors has proved successful for isolation of dominant and/or recessive mutations both in genes that were previously known *(fem-1, fem-2, fem-3, tra-1, tra-2)* and in novel genes [*fog-1, fog-2, sup(q62), sup(q80), sup(q81)*] (Kimble et al. 1986; Kimble laboratory, unpublished results).

In addition to the use of genetic selections, we have conducted simple mutagenesis screens to isolate mutations that sexually transform the germ line. These screens have been done to complement the genetic selections described above in an attempt to identify all the genes that can mutate to a Fog or Mog phenotype. The progeny of approximately 8000 F1 clones have been examined for mutations that cause sexual transformation of the germ line (Maples and Kimble, unpublished). Mutations in several *fog* and *mog* genes have been isolated by this screening procedure: *fog-1(I), fog-2(V), fog(q157)II, fog(q163)III, mog-1(III)*, and *mog-2(II)*.

3.4.2 Global Sex Determination Genes with Germ Line-Specific Controls

Gain-of-function alleles of two global sex determination genes, *fem-3* (Barton et al. 1986) and *tra-2* (Doniach 1986), cause sexual transformation of the hermaphrodite germ line without affecting its soma. Neither affects the XO male sexual phenotype. A *fem-3(gf)* hermaphrodite makes sperm but no oocytes and is therefore Mog; a *tra-2(gf)* hermaphrodite makes oocytes but no sperm and is therefore female. Loss-of-function alleles of either *fem-3* or *tra-2* cause sexual transformation of both soma and germ line.

The activity of *fem-3* appears to be under at least two controls in wild-type hermaphrodites (Barton et al. 1987). First, in wild-type hermaphrodites, the influence of *fem-3* is limited to the germ line; spermatogenesis, a *fem-3*-dependent male process, occurs without masculinization of the female (hermaphrodite) soma. In *fem(gf)* hermaphrodites, like in wild-type hermaphrodites, mesculinization is confined to the germ line. One simple mechanism for this tissue-specific limitation might be that *fem-3* is produced in the germ line and not in the soma. Alternative possibilities include the activation of *fem-3* late in development, at a time when the somatic sex has been determined, a lower threshold for action of *fem-3* in the germ line compared to the soma, or tissue-specific regulation of another gene (or its product) upon which the activity of *fem-3* depends.

Second, in wild-type hermaphrodites, although the activity of *fem-3* is necessary for spermatogenesis, it must be regulated to permit the onset of oogenesis. In *fem-3(gf)* hermaphrodites, sperm are made continuously and in vast excess (Barton et al. 1987). A likely explanation of the defect resulting in this mutant phenotype is that the *fem-3(gf)* gene, or its product, is no longer sensitive to a negative control that is normally imposed to begin oogenesis. A negative regulation of *fem-3* activity is also implied by the maternal effects observed for *fem-3(lf)* mutants (Hodgkin 1986; Barton et al. 1986). If *fem-3* is produced in oocytes for sex determination in the embryo, then the product packaged in the oocyte must be in an inactive form, otherwise, the oocyte might have been transformed into a sperm. The inactivation of *fem-3* in the oocyte must be reversible if *fem-3* is to function during embryogenesis. Thus, negative regulation of *fem-3* activity is suggested both to permit the onset of oogenesis and to permit packaging of *fem-3* into oocytes.

The wild-type activity of *tra-2* is required for both somatic development and the production of oocytes in hermaphrodites (Hodgkin 1980). Thus, in XX animals, *tra-2* influences the sexual phenotype of all tissues. However, since the wild-type *tra-2* gene feminizes the germ line, that activity must be modulated to permit spermatogenesis in hermaphrodites. In *tra-2(gf)* hermaphrodites, no sperm are produced (Doniach 1986). This defect may be caused by a loss of sensitivity of the *tra-2(gf)* gene, or its product, to a negative control that is normally imposed to allow spermatogenesis (Doniach 1986). As described below, a good candidate for the negative regulator of *tra-2* is *fog-2* (Schedl and Kimble, manuscript in preparation).

3.4.3 Germ Line-Specific Sex Determination Genes

Mutations in *fog* or *mog* genes result in a transformation of cell fate. Germ cells that would normally undergo spermatogenesis in wild-type animals undergo oogenesis in *fog* mutants. Similarly, germ cells that would normally undergo oogenesis in wild-type animals undergo spermatogenesis in *mog* mutants. In XX animals, *fog* and *fem* mutants are identical. However, in XO animals, *fog* mutants are never feminized in the soma, whereas *fem* mutants are feminized in both soma and germ line. Therefore, the Fog phenotype is distinct from the Fem phenotype.

One *fog* gene, *fog-2,* has been characterized more than the others (Schedl and Kimble, manuscript in preparation). XX animals homozygous for any of eleven mutations in *fog-2* are female, whereas XO animals are male. The *fog-2* alleles are all recessive. Though none is an amber allele, they arise after EMS mutagenesis at a frequency typical of loss-of-function mutations in other genes, and they all have the same mutant phenotype. This suggests that these mutations cause a loss of the *fog-2* product.

The *fog-2* gene is not required for spermatogenesis *per se* since male spermatogenesis is not affected. Instead, *fog-2* is required for hermaphrodite spermatogenesis specifically. To determine how *fog-2* influences spermatogenesis in hermaphrodites, strains mutant at both *fog-2* and one of the other sex determination loci have been examined. The *tra-2;fog-2* double-mutant phenotype is identical to the *tra-2* null phenotype: sperm are produced continuously in the germ line (Schedl and Kimble, manuscript in preparation). The *fog-2;her-1* double-mutant phenotype is identical in XO animals to the mutant phenotype of *fog-2* in XX animals: XO hermaphrodites make no sperm. These experiments suggest that *fog-2* may be a negative regulator of *tra-2*, and provide evidence that *fog-2* is required for the onset of spermatogenesis in hermaphrodites, whether XX or XO. The gain-of-function phenotype of *tra-2* is identical to the loss-of-function phenotype of *fog-2*. If *fog-2* negatively regulates *tra-2*, then *tra-2(gf)* may be defective in regulation by *fog-2*. This possibility predicts that *tra-2(gf)* and *fog-2(lf)* mutations should behave identically in genetic tests and that the effects of *tra-2(gf)* and *fog-2(lf)* mutations should not be additive. These predictions are currently being tested.

Two other *fog* genes, *fog-1* (Hodgkin et al. 1985; Barton and Kimble, unpublished) and *fog(q126)* (Schedl and Kimble, unpublished), are required for spermatogenesis in both XX hermaphrodites and XO males. Mutation of either gene causes feminization of the germ line of both hermaphrodites and males. The null phenotype of neither *fog-1* nor *fog(q126)* has been established. Recently, numerous alleles of *fog-1* have been isolated as dominant suppressors of *fem-3(gf)* (Barton and Kimble, unpublished). The rapid isolation of many alleles of this gene should therefore provide information on its null phenotype.

Mutation of either of two *mog* genes results in the production of excess sperm with no signs of oogenesis in the XX hermaphrodite gonad (Schedl and Kimble, unpublished results). The *mog* genes have no effect on the sexual phenotype of the somatic tissue of either XX or XO animals. Unlike *fog* mutants, which make oocytes and therefore are cross-fertile, homozygous *mog* mutants are self- and cross-sterile. Genetically, the *mog* mutants are therefore more difficult to manipulate and have lagged behind the *fog* mutants in their characterization.

4 Conclusions

The exact lineal descent of a germ cell does not direct its differentiation as a sperm or oocyte in the nematode, *C. elegans*. Similarly, neither chromosomal sex nor influences from somatic tissues appear to direct the decision between spermatogenesis and oogenesis. However, two classes of genes are involved in this control. The global sex determination genes affect the sexual phenotype in all tissues, including both somatic and germ line tissues. Two of the global sex determination genes (*fem-3* and *tra-2*) appear to be specifically regulated to achieve the switch from spermatogenesis to oogenesis typical of hermaphrodites. In addition, there are at least five germ line-specific sex determination genes that affect the decision between spermatogenesis and oogenesis specifically.

Acknowledgments. This work was supported by NIH through grants GM31816 and HD00630, and by the March of Dimes Birth Defects Foundation by Basil O'Connor Starter Research Grant No. 5-514.

References

Barton MK, Schedl TB, Kimble J (1987) Gain-of-function mutations of *fem-3*, a sex determination gene in *Caenorhabditis elegans*. Genetics 115:107–119
Brenner S (1974) The genetics of *Caenorhabditis elegans*. Genetics 77:71–94
Doniach T (1986) Activity of the sex-determining gene *tra-2* is modulated to allow spermatogenesis in the *C. elegans* hermaphrodite. Genetics 114:53–76
Doniach T, Hodgkin J (1984) A sex-determining gene, *fem-1*, required for both male and hermaphrodite development in *Caenorhabditis elegans*. Dev Biol 106:223–235
Eide D, Anderson P (1985) Transposition of Tc1 in the nematode, *Caenorhabditis elegans*. Proc Natl Acad Sci USA 82:1756–1760
Greenwald I (1985) *lin-12*, a nematode homeotic gene, is homologous to a set of mammalian proteins that includes epidermal growth factor. Cell 43:583–590
Herman RK (1984) Analysis of genetic mosaics of the nematode *Caenorhabditis elegans*. Genetics 108:165–180
Hirsh D, Oppenheim D, Klass MK (1976) Development of the reproductive system of *Caenorhabditis elegans*. Dev Biol 49:200–219
Hodgkin J (1980) More sex-determination mutants of *Caenorhabditis elegans*. Genetics 96:649–664
Hodgkin J (1983a) Two types of sex-determination in a nematode. Nature 304:267–269
Hodgkin J (1983b) Switch genes and sex determination in the nematode *C. elegans*. J. Embryol Exp Morphol 33:103–117
Hodgkin J, Doniach T, Shen M (1985) The sex-determination pathway in the nematode *Caenorhabditis elegans*: variations on a theme. Cold Spring Harbor Symp Quant Biol 50:585–593
Hodgkin J (1986) Sex determination in the nematode *C. elegans*: Analysis of *tra-3* suppressors and characterization of *fem* genes. Genetics 114:15–52
Hodgkin JA, Brenner S (1977) Mutations causing transformation of sexual phenotype in the nematode *Caenorhabditis elegans*. Genetics 86:275–287
Kimble J, Hirsh D (1979) Post-embryonic cell lineages of the hermaphrodite and male gonads in *Caenorhabditis elegans*. Dev Biol 87:286–300
Kimble J, Schedl T (1987) Developmental Genetics of *Caenorhabditis elegans*. In: Malacinski GM (ed) Developmental Genetics, A Primer in Developmental Biology. Macmillan New York
Kimble J, Sharrock WJ (1983) Tissue-specific synthesis of yolk proteins in *C. elegans*. Dev Biol 96:393–402
Kimble J, Ward S (1987) Germline development and fertilization. In: Wood WB (ed) The biology of *Caenorhabditis elegans*. Cold Spring Harbor Press
Kimble J, White JG (1981) On the control of germ cell development in *Caenorhabditis elegans*. Dev Biol 81:208–219
Kimble J, Hodgkin J, Smith T, Smith J (1982) Suppression of an amber mutation by microinjection of suppressor tRNA in *C. elegans*. Nature 299:456–458

Kimble J, Edgar L, Hirsh D (1984) Specification of male development in *Caenorhabditis elegans*. Dev Biol 105:234–239

Kimble J, Barton MK, Schedl TB, Rosenquist TA, Austin J (1986) Controls of postembryonic germ line development in *Caenorhabditis elegans*. In: Gall JG (ed) Gametogenesis and the early embryo, 44th Symp Soc Dev Biol. Alan R Liss, New York

Klass MK, Wolf N, Hirsh D (1976) Development of the male reproductive system and sexual transformation in the nematode *Caenorhabditis elegans*. Dev Biol 52:1–18

Madl JE, Herman RK (1978) Polyploids and sex determination in *Caenorhabditis elegans*. Genetics 93:393–402

Moerman DG, Waterston RH (1984) Spontaneous unstable *unc-22 IV* mutations. Genetics 108:859–877

Nelson GA, Lew KK, Ward S (1978) Intersex, a temperature-sensitive mutant of the nematode *Caenorhabditis elegans*. Dev Biol 66:386–409

Sulston JE, Brenner S (1974) The DNA of *Caenorhabditis elegans*. Genetics 77:95–104

Sulston JE, Albertson DG, Thomson JN (1980) The *Caenorhabditis elegans* male: postembryonic development of nongonadal structures. Dev Biol 78:542–576

Sulston JE, Schierenberg E, White JG, Thomson JN (1983) The embryonic lineage of *Caenorhabditis elegans*. Dev Biol 100:64–119

Trent CN, Tsung N, Horvitz HR (1983) Egg-laying defective mutants of the nematode *Caenorhabditis elegans*. Genetics 104:619–647

Wills N, Gesteland RF, Karn J, Barnett L, Bolten S, Waterston RH (1983) Transfer RNA-mediated suppression of nonsense mutations in *C. elegans*. Cell 33:575–583

Subject Index

abnormal spermatozoon head (azh) 6
acrosome, absence 6, 13
actin 46
albino (c) 7
allocyclic behaviour of sex chromosomes 34, 37
androgen receptor 21
autonomous gene expression in germ cells 21, 25, 40
axonemal defects 98
axoneme 4, 99
azoospermia 6, 31

blind-sterile (bs) 6
blood-testis barrier 3
bobbed (bb) 68

C-banding, of Y chromosome 78
cell lineage
  germ line 119
  in C. elegans 117
centriole 65
chimeric mice 25
chromatin, structural alterations 14, 105
chromosomal aberrations, sterility of 23, 24, 27
collochore 68, 101
crossing over, meiotic 69
cyst cell 63
cytoplasmic bridges (see intercellular bridges)

distorter genes (Tcd) 15, 19
dosage compensation 68
dynein 100

epididymis 5
exchanges, gonial 65

feminization 120, 126
fertility, quantitative effects 107
fertility genes
  intragenic complementation 71
  Y chromosomal 69
flagellum 65

G-banding, of Y chromosome 92
germ cell, chromosomal sex and differentiation 123
germ cell depletion 6
germ cell line, committment 122
germ cells development, control genes 124
germ line, sexual transformation 124
Giemsa staining, of Y chromosome 90
Golgi apparatus 4

H–Y surface antigen 39, 40
Haldane's rule 34
haploid gene action 45, 51
head cyst cell 65
hermaphrodite 117, 121
Hoechst 33528-banding, of Y chromosome 78, 90, 92
hop-sterile (hop) 7
hybrid sterility-1 (Hst-1) 8

imprinting, paternal 39
individualization 64, 65, 97, 98
intercellular bridges 1, 48, 49, 63, 65
intervening sequence of rDNA (IVS) 89

lambrush loop 63, 103, 106
  species specificity 93
lampbrush loops 83
  transcripts 96
lampbrush loop structure, autonomy 85

male fertility, autosome aberrations 28
male sterile mutations 5, 66
male sterility, in X–A translocations 24, 67
manchette 19
meiotic drive 65
meiotic recombination, of Y chromosome 77
microtubules 19, 97, 99, 106
middle piece 4
mitochondrium 4
mosaics, genetic 119
mutations, unstable 75

Subject Index

N-banding, of Y chromosome 78, 81, 83, 90
nebenkern 65, 97, 100
noncomplementors, second-site 106
nucleolus organizer region (NOR) 68
nucleus, elongation 4

oligotriche (olt) 9
oncogene c–abl 47
ovotestis 118, 119

p alleles 9
pairing, homologous chromosomes 3
pairing sites, and fertility 31
paracrystalline material 65, 98
PGK-2 47
position effect variegation 68, 69, 89
progenitor cell P4 122
protamine 44, 47, 48, 50
pseudopod, of spermatozoon 120
Purkinje cell degeneration (pcd) 10

quaking (qk) 11
quinacrine staining, of Y chromosome 78, 90

responder genes (Tcr) 15, 19
ribosomal DNA 67, 68
ring channel 63

satellite DNA, in Y chromosome 83
segmental aneuploidy 73, 81, 99, 107
segregation distortion 65
seminal vesicle 65
Sertoli cells 1, 3, 40
sex chromosome, pairing 26
sex determination
  C. elegans 120
  genetic regulation 121
  in germ line 122
sex reversed (Sxr) 23, 37, 40, 41
sheat, fibrous 4
sperm degeneration 98
sperm dimensions, differences in mice 12
sperm head, abnormal 6, 13
sperm motility 18
sperm tail
  abnormal 13
  absence 8
spermatocytes
  primary 1–3, 63
  secondary 3, 65
spermatogenesis factor 41

spermatogonia 2, 41
spermiogenesis
  Drosophila 64
  mammalian 3
stellate locus (Ste) 77, 101, 105, 108
stem cell 2, 63
sterility, and structural organization of the genome 108
sterility by Y chromosome triplications 77
suppressor mutations 118
synaptonemal complex 26
synthetic sterility 86

t complex 14
t complex peptide 1 (TCP–1) 47
t haplotypes 12, 14, 18, 19, 22
TAG staining of Y chromosome 90
TCP–1 (see t complex polypeptide)
Tc1 (see transposable element) 119
testicular feminization (Tfm) 21
testis-determining gene 39
  (Tdy) 40
transcription, during meiosis 42, 44
transmission rate distortion (see distorter genes and meiotic drive)
transposable element 104, 119
tubules, seminiferous 1, 2
tubulin, testis-specific 8, 46, 48, 102, 106

underreplication, of Y chromosome 78

vas deferens 66

wobbler (wr) 12

X chromosome, heteropyknotic morphology 33
X chromosome inactivation 32, 34, 38, 39, 67, 87, 89, 108
X inactivation, time of 38
XY body 35

Y chromosome
  DNA of 83, 103
  function in spermiogenesis 34, 39, 63, 67, 68
  genetic map 70
  influence on sperm shape 12
  structure in Drosophila 84

zygotene DNA 42